Technical Writing Simplified

The simple yet effective, 7-step Technical Writing Process that anyone can use to write documents such as user guides, research papers, and procedures to excel professionally

KARIM PANJWANI

Copyright © 2024 Karim Panjwani

All rights reserved.

Disclaimer

The publisher and author have applied their utmost efforts in the preparation of this book, which includes hardcopy, e-book versions, and any other formats. However, they do not provide any assurances, or guarantees regarding the accuracy or completeness of the contents of this book. They expressly disclaim any implied warranties of merchantability or suitability for a specific purpose. This book is not intended to serve as legal advice, it is provided only as a general guide. The advice, information, and templates contained within this book may not be appropriate for your situation. Some of the advice, information, or templates in this book may become outdated due to changes in industry practices, technology, or law. Neither the publisher nor the author will be held liable for any loss or other damages that may be encountered as a result of using the content of this book or acting on the advice or information contained in this book.

Trademarks

The diagram of 'Technical Writing Simplified' is a trademark of Karim Panjwani. Adobe, Acrobat, FrameMaker, InDesign, and Photoshop are either registered trademarks or trademarks of Adobe Systems Incorporated in the United States and/or other countries. Apache and OpenOffice are trademarks of The Apache Software Foundation. Author-it is a registered trademark of Author-it Software Corporation. BPMN is a trademark of Object Management Group, Inc. in the United States and/or other countries. ISO is a registered trademark of the International Organization for Standardization. Microsoft, Excel, PowerPoint, SharePoint, and Visio are either registered trademarks or trademarks of Microsoft Corporation in the United States and/or other countries. TechSmith and Snagit are registered trademarks of TechSmith Corporation. All other trademarks and registered trademarks specified in this book mentioned above or anywhere else are the property of their respective owners.

Acknowledgment

This book is the result of years of learning. I couldn't write this book without the support of my family, friends, mentors, and colleagues.

Firstly, I owe gratitude to my mentors from my educational and professional careers. Your guidance and teaching have been precious in shaping my understanding and what I am now. Without your teaching and guidance, I could not write this book.

To my peers in the industry, your companionship and shared experiences have enriched my perspective and added depth to this work. The knowledge I gained from peers, whether through formal interactions or casual conversations, has been instrumental in shaping this book.

I extend my heartfelt appreciation to all my industry colleagues who have interacted with me over the years. Your insights, experiences, and wisdom have been a constant source of inspiration. This book is a testament to our collective knowledge.

A special note of thanks to the tireless editors, researchers, and professionals who reviewed and provided valuable feedback on early drafts. Your meticulous attention to detail and constructive insights have significantly enhanced the quality of this work.

Last, but certainly not least, I am grateful to my family and friends for their unwavering support and understanding during the writing process. Your encouragement has been a source of motivation, and your belief in the importance of this endeavor has been my driving force.

In conclusion, "Technical Writing Simplified" is not just my creation, but a product of the collective wisdom and efforts of many. To everyone who has played a part, no matter how small, in bringing this book to fruition, I extend my deepest gratitude. Thank you.

CONTENTS

Chapter 1: Introduction   1

Chapter 2: Technical Writing Process   5

Chapter 3: Objective   11

Chapter 4: Plan   18

Chapter 5: Framework   36

Chapter 6: Draft Writing   47

Chapter 7: Review   92

Chapter 8: Edit   105

Chapter 9: Release   113

Chapter 10: Conclusion   121

A Request to The Reader   124

References   125

About the Author   127

Chapter 1: Introduction

Clarity in technical writing is not just a skill; it's a gateway to innovation.

'If you can't explain it simply, you don't understand it well enough.'

~ Albert Einstein

Welcome to the world of Technical Writing! This book serves as your guide, illuminating the path to mastering this invaluable skill. It's not just for those in technical professions - anyone can benefit from honing their technical writing abilities. This book is your companion, helping you navigate the content and apply it to your unique context.

I've distilled the essence of technical writing into a straightforward, seven-step process. This book presents the entire process in a simplified format, breaking it down into typical steps that are easy to follow.

Step by step, this book will guide you in developing and enhancing your technical writing skills. It's like having a mentor by your side, sharing insights and advice from over 17 years of experience in the technical profession.

This book is more than just a guide - it's a tool, designed from personal experience and valuable advice gathered throughout my career. I hope that this book will assist you in planning, drafting, and designing documents that will make a significant impact on your professional life.

Embark on this journey of technical writing with us. Good luck, and may this book serve as your beacon, guiding you toward becoming an accomplished technical writer!

Who can be benefited with this book?

This book is a valuable resource for students and professionals who engage in technical writing. This includes engineers, scientists, pharmacists, academic book writers, architects, physicians, lab technicians, office documentation staff, and many others. In fact, most professions require at least occasional technical writing. Examples of such writing include scientific experiments, scientific journals, responses to customer complaints and legal inquiries, business communications, user guides, manuals, and investigations for event reports in the organizations during manufacturing processes.

Creating good technical documentation is not a spontaneous process. It requires the production of content that is both technically accurate and user-friendly - a highly specific skill.

This book can be a helpful guide for individuals associated with any of the above professions. Many companies have Subject Matter Experts (SMEs), but they may lack a professional writer on staff who can translate complex product knowledge into content that's easily understood by the end-user. Many companies also lack the internal resources to meet technical documentation demands, leading them to outsource this function to a technical writing expert.

This book can be used by a wide range of audiences who want to enhance their technical writing skills, those who find themselves tasked with technical documentation, technical writers who want to bring more structure to their work, and anyone managing documentation work.

While hard skills are essential to even getting a foot in the door of a company, a repository of strong tangible skills does not guarantee speedy growth for anyone. Whether you're working in a growing startup or an established MNC, there are a few essential skills that must be developed and honed if you want to progress in your career.

What could go wrong?

Precision and accuracy are the cornerstones of technical writing. Imagine the catastrophic consequences of an error in the launch process of a missile. The repercussions of poor communication in business and industry can be severe, leading to loss of business, projects, customers, and even legal complications.

Inefficient emails, careless reading or listening to instructions, unread documents due to poor design, hastily presented inaccurate information, and sloppy proofreading - all these result in inevitable costs. These losses can be measured in wasted time, work, money, and ultimately, professional recognition. In extreme cases, losses can even be measured in property damage, injuries, and deaths.

Consideration while writing this book

This book is designed with the practical aspects of document development in mind. The process of technical writing is broken down into seven steps - Objective, Plan, Framework, Draft Writing, Review, Edit, and Release. Each step is explained in a separate chapter with practical examples.

The book is designed for those who want to systematically understand the process of technical writing, such as beginners, managers, and subject matter experts. The content of the book is kept as simple as possible for easy understanding.

Each chapter provides definitions, useful tips, notes, and practical examples as applicable to help readers easily understand the text. This book is your guide to mastering the art of technical writing. Are you excited? Let's embark on this journey together!

Symbol	Explanation
DEFINITION	Definition highlights are simple explanations that clarify complex technical terms used in the text.
QUICK TIPS!	Quick Tips are practical suggestions that assist in understanding and applying the ideas discussed in a specific section.
Notes	Notes provide supplementary details about a topic or section that, while not crucial, can enhance the reader's understanding of the subject.

Chapter 2: Technical Writing Process

"Simplicity is the ultimate sophistication."

~ Leonardo da Vinci

Welcome to the second chapter of our journey into the realm of Technical Writing. This chapter provides a brief overview of the importance of Technical Writing skills and outlines who can benefit the most from acquiring these skills. It also offers guidance on how to best utilize this book and comprehend its content.

You might be wondering, is there a difference between technical communication and technical writing? Well, they are two sides of the same coin. Communication can be achieved through various methods, and writing is one of them. In this sense, a technical writer is indeed a technical communicator. Traditional communication methods included writing manuals, books, letters, or literature. Today, technical writers also create content such as presentations for forums, video training materials, technical charts, and any other form where technical knowledge can be translated for the audience in a simple and easy-to-understand manner.

Methodology

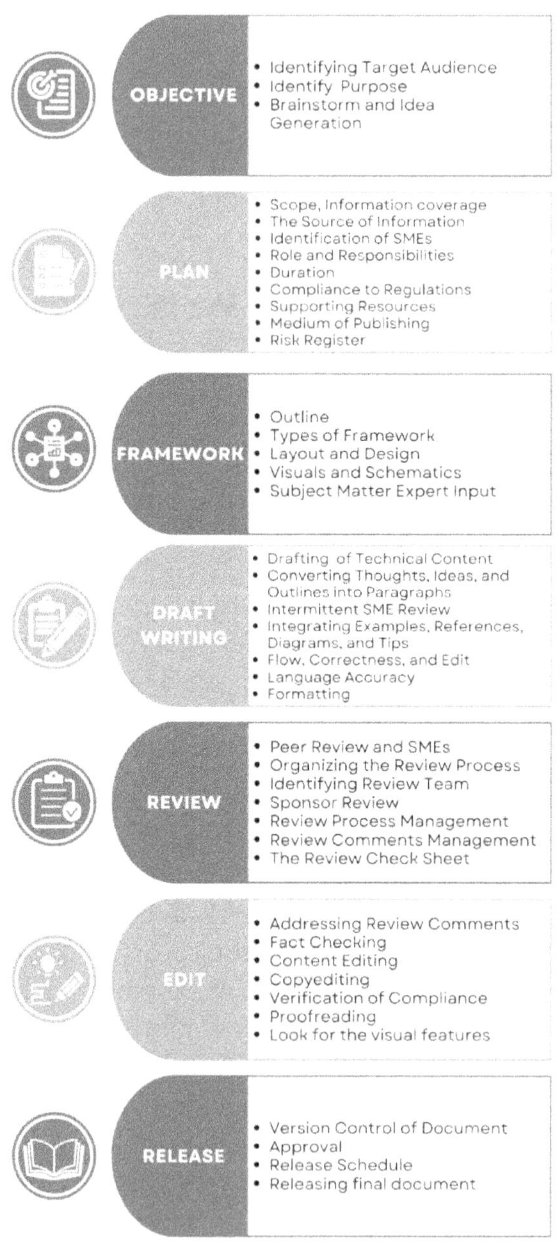

At the heart of the technical writing process is a seven-step approach, and the structure is elucidated through infographics.

This overarching process is divided into seven key steps to create a technical document - Objective, Plan, Framework, Draft Writing, Review, Edit, and Release. As you delve deeper into the book, you'll find that this high-level structured approach applies to most documentation work.

The methodology defined in this book will equip you with a skill, enabling you to break down complex processes into easy-to-understand write-ups. This book provides a systematic process for writing and developing content that meets professional standards.

The beauty of breaking down the process into smaller steps lies in its ability to simplify complex information innovatively. This structured approach transforms the process into a team task, eliminating duplicated efforts. It brings a sense of certainty when a writer embarks on crafting a technical document. The process also aids in determining the role, responsibility, and accountability at each step. It's important to note that there are no fixed or definite rules for these steps and their chronology. The steps are interdependent and overlapping, meaning we might revisit previous stages in the process from time to time. Depending on the complexity of the information and the situation, steps can be added or removed from the process.

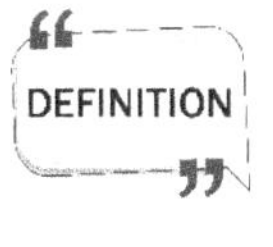	*Structured Approach:* In terms of technical writing, the Structured Approach refers to a systematic and organized way of creating content to make the process of technical writing easy and timebound.

Application of Technical Writing

Since ancient times, technical writers have served as a bridge between inventors and users. Technical writing is the process of transforming technical knowledge into simple language to aid a specific audience's understanding. The term 'technical' has a broad meaning. It refers to knowledge relating to a particular subject, art, or craft, or its techniques, considered technical for that particular specialization.

The beauty of being a technical writer is that they are needed in almost any industry. This includes the pharmaceutical industry, research organizations, the software industry, the food industry, law firms, various manufacturing industries, the infrastructure industry, and many more.

Industries demand a variety of documentation from technical writers. The types of documents in technical writing can be broadly categorized into four categories (shown in Figure 1).

Figure 1: Types of technical writing

<table>
<tr>
<td>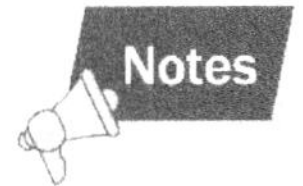
</td>
<td>What do you understand when you read the term "Technical"?

Generally, when someone uses the term 'technical', our minds immediately think about fields like engineering or science. However, the word 'technical' has a broader perspective. It refers to any knowledge that is specific to a particular art, subject, or craft, and the specialized knowledge associated with it. This knowledge is usually gained through study or experience in that field.

For instance, a subject matter expert in the field of commerce who has in-depth knowledge of financial markets, an artist who has mastered the techniques of a specific art form, or a manager who knows the ins and outs of management of his or her area - all possess 'technical' knowledge in their respective fields.</td>
</tr>
</table>

Advantages of the Structured Process of Technical Writing

The process of structured writing offers numerous benefits to both users and writers.

Time-bound: Writers can complete tasks with high productivity and efficiency. The objectives of the document, defining an audience, and planning the project are well-understood at the early phase of the process. This process also allows for collaboration with SMEs and stakeholders, setting expectations from the beginning of the project. This helps writers finalize the content of the document faster, saving a significant amount of time and cost for an organization.

Reader-friendly: The focal point is the end user. Considering the target readers is the key to meeting expectations. This helps writers select the content, depth of information, word choice, and tools for explanation such as diagrams, schematics, pictures, tables, statistics, or simple examples. Readers will receive information quickly, to the point, and in easy-to-understand language.

Accuracy and Consistency: The book provides a structured process to create a document. This helps in reducing the potential for errors, missing information, misleading content, and inconsistencies.

Lifelong Learning for Writers: One of the positive aspects of the process is that writers can learn by involving themselves in the process, and collaborating with the team of SMEs and stakeholders. To effectively communicate the information to the readers, writers must understand the core knowledge.

Useful for Beginners, Students, and Professionals: The structured approach defined in the book is a helpful tool for those who are thinking of becoming a technical writer, those who are already in the field and want to improve their skills to the next level, those whose job demands the task of creating technical documentation, writers who want to further organize their approach to writing documents, and anyone in the field of documentation projects.

Reputation for Individuals and Organizations: A superior, skillfully written document, literature, and manuals reflect well on your organization and the individual who created the document. Technical writers help translate jargon into simpler language and meet the needs of the target audience.

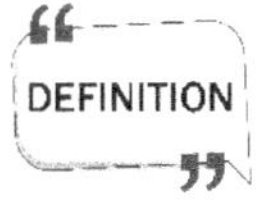

Subject Matter Expert (SME):	A Subject Matter Expert (SME) is a person who has a significant understanding and expertise in a specific field or area.

Subject Matter Expert (SME):

A Subject Matter Expert (SME) is a person who has a significant understanding and expertise in a specific field or area.

Stakeholders:

In simple terms, any person or group who is positively or negatively impacted by an activity, initiative, policy, or organization. It comprises members, such as employees, customers, communities, suppliers, associations, and government agencies.

Jargon:

Jargon means special or technical terms that are used by a particular group of professionals in a specific profession that other people do not understand.

Chapter 3: Objective

"Setting goals is the first step in turning the invisible into the visible."

~ Tony Robbins

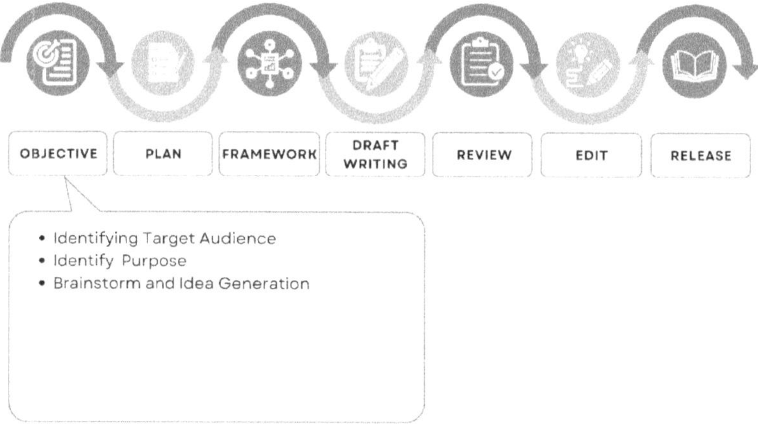

The creation of a technical document is driven by various objectives. Understanding these objectives is a crucial first step in technical writing. A clear objective provides the writer with a direction, helping to shape many aspects of the writing process and charting a path forward.

Identifying the Target Audience

When you're writing any document, your goal is to convey a message to your readers. The aim of writing is to communicate ideas precisely. Taking the readers' needs into account is crucial to achieving your purpose. The same technical information may need to be conveyed differently for different groups of audiences.

The target audience for a user manual is typically the general public, and their level of understanding can vary depending on factors such as educational background, region, age, etc. The content of the document might include unboxing the product, assembling different parts (if any), plug-and-play instructions, and troubleshooting. The content might need to be in multiple languages, require pictorials as well as symbols, and be written in very simple language. In such cases, an additional step might be required in the writing methodology, which is the translation of content into different languages.

Similarly, if the document to be created is for patent filing for scientific innovation, the content of technical writing would demand highly technical language. It would also require a legal perspective and need to be written in a specific format for the submission purpose to get the filing accepted by authorities. The stakeholders and SMEs who might need to be involved in the project could be the product development team, legal team, and due diligence team.

A simple yet effective tool to understand the audience's requirements is the 5W 2H tool. This tool is straightforward but very powerful in determining the entire text of technical writing. 5W 2H stands for Who, What, When, Where, Why, How, and How much.

- **Who:** This refers to the readers of the document. Understanding who the readers are is crucial as it helps tailor the content to meet their needs and level of understanding.

- **What:** This refers to the use of the document for the readers and what kind of text they expect. It's about understanding the readers' needs and expectations from the document.

- **When:** This refers to the time when the reader will use the document. Knowing when the document will be used can help in structuring the content and making it relevant for that time.

- **Where:** This refers to the location where the document will be used. The location can influence the language, tone, and even the format of the document.

- **Why:** This refers to the reason why the user needs the document. Understanding the 'why' can help in making the content more purposeful and relevant.

- **How:** This refers to how the user will use the document and how the reader is going to read the document. It can influence the organization of the content, the level of detail, and the inclusion of elements like summaries, tables of contents, indexes, etc.

- **How much:** This refers to the amount of content that could be useful for the targeted audience. It's about striking the right balance between providing enough detail to be informative, but not so much that it overwhelms the reader.

By answering these questions, you can ensure that your technical document is well-tailored to your audience's needs and expectations. It's a valuable tool in the technical writing process.

Once you have all the answers to these questions, your primary outline is ready, and you can proceed to the next step of the technical writing process.

Identify the Need and Establish Primary Purpose

This is the pre-writing phase of creating a technical document. No technical document is created without a purpose. The project starts when a technical document is requested by an employer, colleague, customer, or when you generate one based on your research or work. Initial requirements are gathered and outlined as the type of document, subject, scope of the document, content to be covered, goal, and audience.

The following are a few examples of the motives for creating technical documents:

i. To prepare a user manual for a general-purpose electronic product

ii. Product promotion literature of highly sophisticated medical equipment

iii. Document for patent filing

iv. Standard operating procedure for equipment operation

v. Instruction for manufacturing a food product or pharmaceutical product

vi. Writing an investigation report for an event that occurred in the pharmaceutical or medical device facility

vii. Supply and Quality Agreement between supplier and customer

By reading the above examples, what thoughts come to your mind from the perspective of a technical writer? Each of the documents has a different target audience and readers. You can determine this using the simple steps of 5W 2H as outlined in the previous section.

Brainstorming and Idea Generation

Brainstorming is a method of idea generation where one person or a team can freely think and generate ideas. Introduced in 1939 by advertising executive Alex Osborn, brainstorming encourages the generation of as many ideas as possible, even wild and exaggerated ones. At this point, no idea is considered best or meaningless; everything that comes to mind is noted down. Brainstorming can stimulate creative and lateral thinking. Once the brainstorming session is completed, all the ideas are reviewed, and duplicates are removed. This process provides a general outline for the intended technical document, helping to finalize the objective of the document. This is a crucial step in the technical writing process, setting the stage for the subsequent steps.

 	It is very important to adopt a user-centric approach. You need to keep a focal point as your audience when writing technical documents. Use the 5W 2H tool to identify the scope of the document, tailoring your content to meet the specific needs, and to meet the expectations of your readers. Understanding your audience is a fundamental aspect of accurately delivering the content to end users.
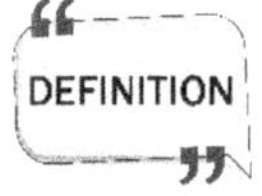 	*Tailored Content:* Tailored content refers to the customization of information that is suitable for the specific needs, understanding level, and absorption capacity of a target audience. The purpose is to make the content as easy as possible for the intended readers.
 	*Summary:* Similar to setting a goal that guides your actions, understanding your audience guides the technical writing. Every document that you create has a purpose and when you design the document as per your audience's specific needs, you are on the right path.

Imagine that you are creating a user manual for a coffee maker. Think - Who are you writing for? A tech-savvy teenager or a senior citizen? The language, format, and even the diagram would be much different! This is where the 5W2H tool comes in handy.

Who, What, When, Where, Why, How, and How much - answering these questions helps you achieve your goal. By keeping readers in mind, you can write a document that is easy to follow, interesting, and goes above and beyond expectations.

Remember, understanding your audience is like setting the foundation for your writing. Once you know your audience, the rest of the writing process becomes easy. So, before you put your pen to paper (or fingers on a keyboard), know your audience.

Chapter 4: Plan

"By failing to prepare, you are preparing to fail."

~ Benjamin Franklin

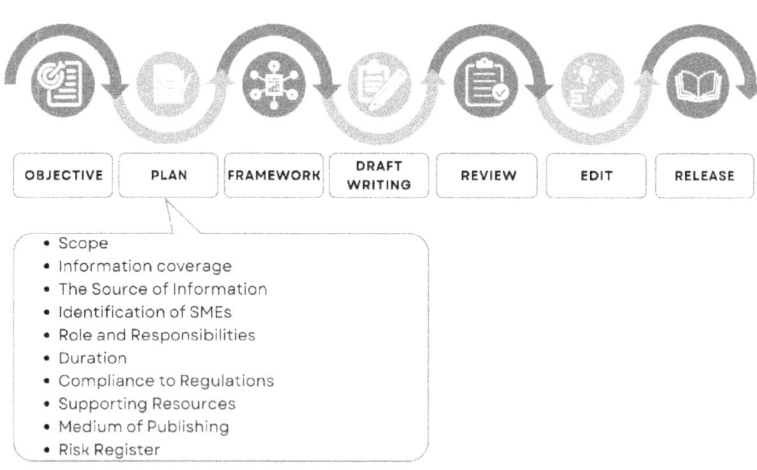

Once you have a clear understanding of your audience, the primary purpose of your document, and a general outline, you can begin to plan your entire project. Remember, planning is the cornerstone of success.

The planning process helps you define the scope of your document, understand your audience, determine the depth of your content, and ensure that the objective of your document is being met. Think of this as your project's roadmap. Developing a plan is a team exercise. The collaboration and consensus of stakeholders are crucial for the timely completion and success of the project. This approach helps eliminate potential surprises and hiccups during the project.

Before you develop a framework for your document and start writing, you need to identify your resources and ensure their availability. These resources include, but are not limited to, information sources, the availability of Subject Matter Experts (SMEs), time points for intermittent Cross Functional Team (CFT) review meetings, the platform to be used for writing the document (e.g., word processor, notebook, etc.), publishing platforms (e.g., email, magazines, scientific journals, etc.), and the requirement of translation into other languages.

Scope of the document

Now that you're equipped with information about your reader, the objective of your document, and the requirements of your audience using the 5W2H method, you can precisely define the scope of your writing. Everything you need to deliver forms the scope of the document. As per the Oxford English Dictionary, meaning of scope is "The sphere or area over which any activity operates or is effective; range of application or of subjects embraced;". In a nutshell, the deliverable sets expectations about what is going to be done, why it's going to be done, and how it's going to be delivered.

Defining the scope is the heart of the document. The simplest way to define the scope is to draw a circle and write down what is in the scope and what is outside the scope. (Refer to the Figure 2). Also, define what the project is not.

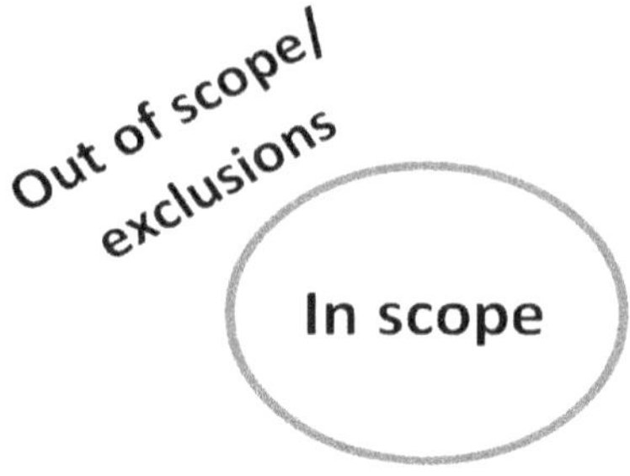

Figure 2: Schematic for way to define the scope

Limit your target audience, establish the limits of functionality, and discuss exclusions. Changing deliverables midway through the writing project could increase the chances of failure. Unnecessarily extending scope boundaries increases the budget, resources, and scheduled time to allow you to deliver.

Identify who the stakeholders are. Stakeholders for the technical document could be engineers, formulators, the marketing team, consultants, and customers. Engage with them and include all the stakeholders in defining the scope. You should not be surprised when you are approaching the end. Be as thorough and detailed as possible, and express deliverables in user's terms. The scope should be tangible and measurable. Once the scope is defined, capture and document it and get approvals from those stakeholders who have the ultimate authority to decide what the final scope will be, and what the final budget in terms of either amount of resources and time will be.

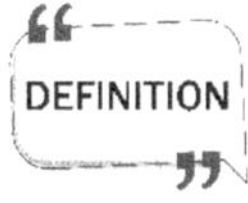

Cornerstone:
Something that is very important or essential. In terms of technical writing, a cornerstone can refer to the most important or fundamental piece of content or information.

Roadmap:
Roadmap refers to a plan or a guide for achieving a certain goal or reaching a certain destination.

Collaboration:
Act of working together or cooperating with others to achieve a common goal.

Consensus:
Agreement or concord among a group of people or the state of being in harmony with someone or something.

Framework:
Framework means a basic structure, plan, system, or set of guidelines that helps in organizing and presenting information effectively.

Cross Functional Team (CFT):
A group of people with different expertise working towards a common objective.

Deliverables:
End output you promise to give or do by the end of a project, that is typically agreed upon by the stakeholders during the initial phase of the project.

Information coverage and depth considering intended audience

In this section of planning, we'll use the information about the audience and the scope of the technical writing to understand the information coverage. The information coverage entirely depends on the identified audience's needs and the scope you defined with the help of a cross-functional team, keeping the objective of the document in mind. List down the topics you are going to cover in the document.

Once the listing of topics has been done, organize it in a logical manner. Mind mapping, a tool invented by Tony Buzan, is an excellent way to organize thoughts in a structured manner. Listing ideas is a linear process; however, the mind does not think in a linear fashion but is an explosion of thoughts. Mind mapping is a visual way to capture thoughts and ideas, making it easier to remember or plan things. To draw a mind map, follow the steps outlined in Figure 3.

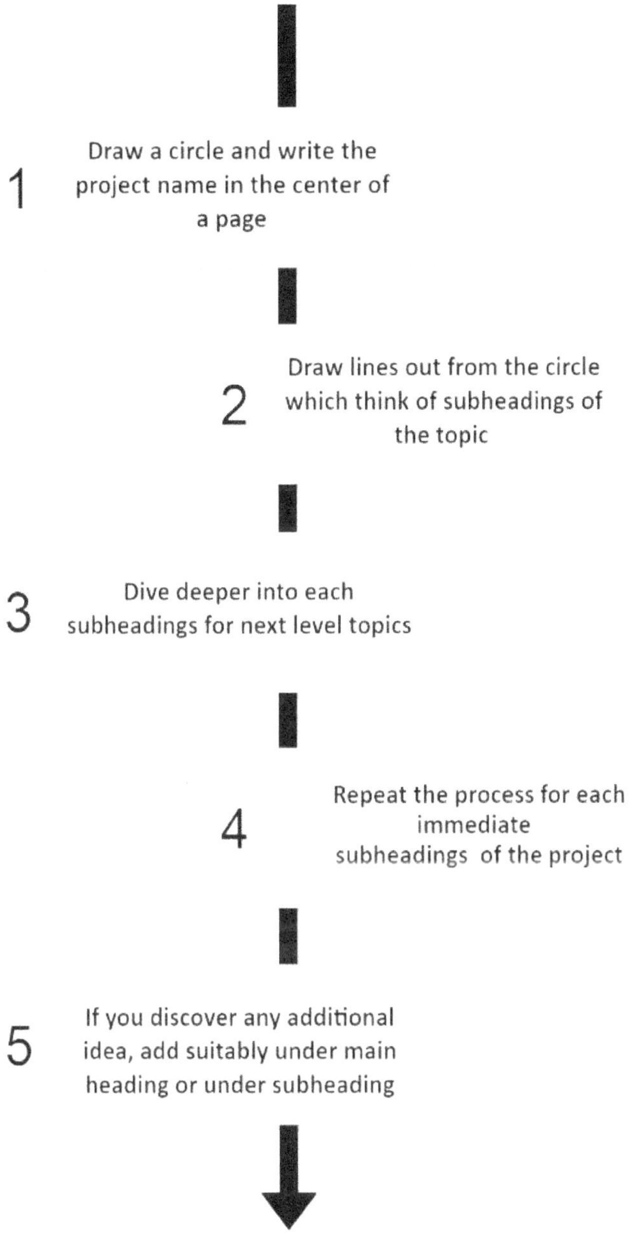

1 Draw a circle and write the project name in the center of a page

2 Draw lines out from the circle which think of subheadings of the topic

3 Dive deeper into each subheadings for next level topics

4 Repeat the process for each immediate subheadings of the project

5 If you discover any additional idea, add suitably under main heading or under subheading

Figure 3: Flow chart of steps to prepare mind map

Suppose you want to write a book on the English language, the mind mapping for the book framework can be done as drawn in Figure 4.

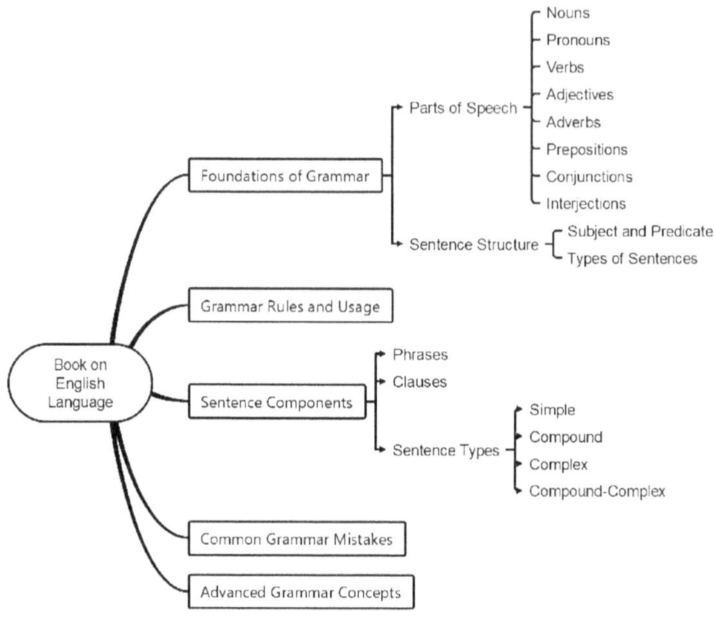

Figure 4: Example mind map to outline the topics for the book on English language

Mind mapping can be done simply using a pen and paper or applications freely available over the internet or purchased from software providers. This will provide a draft table of content for your technical document.

Once the topics are organized in a logical manner and the framework is ready, you should outline the depth of the information needed based on the audience group for whom the document is intended. For example, if you are writing a topic on anatomy and physiology for medical students, you need to provide all the information with a detailed explanation. However, if you are writing the same topic for students of any paramedical branch, such as nursing students, the depth of the content would be lighter than for medical students. Similarly, if you are writing a topic on the electrical motor for electrical engineering students, the topic would cover all possible information starting from the principle of working, mathematical equations to troubleshooting. Whereas, when you are writing the same topic for a student for vocational education, the content would be limited to the basic principle, repairing, troubleshooting, etc. You might also need to write the topic in a vernacular language.

The steps of the technical writing process may require customization depending on project complexity and to fulfil the expectations of the targeted audience. For example, a translation step needs to be added in the process of technical writing where content is needed in multiple languages.

The Source of Information

To craft any technical document, accurate and reliable information is indispensable. This information can be gathered from a variety of sources such as meeting notes, literature, articles, product development reports, the internet, libraries, interviews with relevant stakeholders, your own observations and knowledge, process walkthroughs, site visits, photographs, diagrams, videos, similar products in the market, your company's websites, technical staff and workers, customers, regulatory bodies, and more. Researching this information will strengthen your content.

The source of information depends on the type of document you are writing. For instance, if you are writing an investigation report for an event in the organization, you would need to review the records of events, photographs of the event, visit the site, interview involved personnel or those who witnessed the event, review the laid down procedures for the work and operations, regulations, history of similar events, facts and figures, and understand the impact of the event on the product, process, and compliance.

If you are writing a standard operating procedure for machine operation, the crucial sources of information would be the machine manual, interviews with experienced users, witnessing the operation carried out by the equipment manufacturer's staff, and those who carried out the installation and training for the staff.

In case you are writing literature to be published or content for a webpage to be published on the internet, ensure that the content is not plagiarized to avoid any legal obligations.

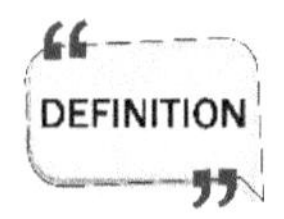

DEFINITION	*Plagiarized or Plagiarization:* Act of using someone else's content and passing them off as your own.

Identification of SMEs

As a technical writer, you may not necessarily know all the technical details; hence, acquiring information from Subject Matter Experts (SMEs) is essential. An SME is an individual with specialized knowledge in a given process or subject. SMEs will provide additional insights on the subject and topic based on their expertise and hands-on experience. This information will offer practical aspects of the subject, which will be very useful for the users of the document.

SMEs could be experienced scientists, equipment operators, customers, marketing executives, consultants, or colleagues with a deep understanding of a process development, function, technology, equipment, material, or particular job. Engaging SMEs intermittently during the process of document writing will help the document creator to speed up the task and eliminate the possibility of failure or delay in the project.

Role and Responsibilities

Defining roles and responsibilities is a critical aspect of completing the task in a timely manner. Usually, the task leader, project leader, or program manager takes the lead in defining roles and responsibilities with the help of a cross-functional team and stakeholders. Role and responsibility go hand in hand with the task schedule. Time and resources are not unlimited for any given task. The writer, task leader/program manager, reviewer, SMEs, and authorities all have roles to play at different stages of technical writing and delivering the planned task as per the predefined timeline. To define the roles and responsibilities, a simple table can be created as follows.

Sr. No.	Task	Responsible stakeholder	Title of stakeholder

The benefit of defining roles and responsibilities at the beginning is that everyone knows the work to be done by them and the timelines.

Project Duration

The time duration of the project is defined as soon as the project starts. The timeline of the tasks is decided either by the client, management, regulatory bodies, or controlled by procedure or policy. Once you define the duration of each set of tasks or task, your workloads will be more controlled and optimized. Task management is a listing of actions and milestones with intended start and completion dates.

Defining a timeline on the schedule will free your mind. Every team member can focus on their task and will effectively work on content, input, and feedback.

Scheduling provides a bird's-eye view of the project path. One can use a simple spreadsheet or a standard project management tool depending on the complexity of the task and the preference of the organization.

The schedule can be prepared using a spreadsheet as follows. The spreadsheet can be converted into a Gantt chart, providing a visual representation of the project timeline. This allows everyone involved in the project to have a clear understanding of the project's progress at any given time. Following is an example of schedule and a Gantt chart that has been created using a spreadsheet.

SUM   X ✓ fx =WORKDAY(D3,C3,G3:G9)

Sr. No.	Task List	No. of days assigned	Start Date	End Date	List of Holidays	Day
1	Project Kick-off	1	03-Jul-20	06-Jul-20	04-Jul-20	Saturday
2	Defining Objective	2	06-Jul-20	=WORKDAY(D3,C3, G3:G9)	05-Jul-20	Sunday
2.1	Establishing primary purpose					
2.2	Identifying audience					
2.3	Brainstorming and idea generation					
3	Plan					
3.1	Scope of the document					

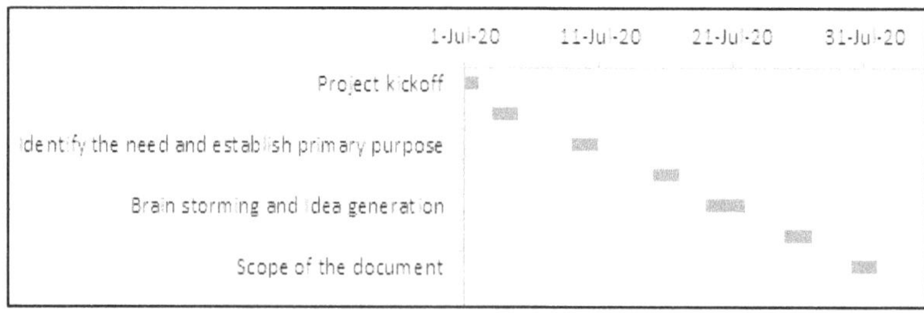

Compliance to Regulations

As writers, we must understand that documents must comply with regulatory standards, regulations, and industry expectations. In medical literature, for instance, the claim for a therapeutic category must be based on clinical studies and approval by a government authority. A technical writer must have supporting information before writing such a claim.

A technical writer also needs to ensure that the content in the document is not copyrighted. Copyright establishes legal protection for intellectual works in printed or electronic form. The copyright owner has an exclusive right to distribute, reproduce, or present work. If a writer plans to reproduce copyrighted material in any publication, permission must be obtained from the copyright holder. A portion of material from a copyrighted source can be utilized for educational use without any permission if it meets the "fair-use" criteria. Even when you use a portion of material that may be reproduced without permission, you must credit the source of information from which the material is taken, appropriately referencing the original author's work.

Intellectual Property:
Creations of the mind, such as inventions, literature, artistic works, designs, or symbols.

Supporting Resources

Having the right resources is crucial for the successful completion of the task. Depending on the type of document to be prepared, different resources are required for technical writers. Examples of various resources include document creation tools, access to libraries, access to the internet, scientific journals, participation in forums and training, image processing tools, drawing creation tools, and graph creation tools. The significance of a few tools and resources are described in the following section.

For document creation, various word processing tools are available, such as Microsoft Word, Apache OpenOffice, FocusWriter, Wordpad, Notepad, Google Docs, and Libre Office. Depending on the business, many organizations subscribe to magazines, e-newsletters, scientific journals. Reading literature helps to acquire information about customer needs, understand the pattern of writing, know the readers' interests, and deal with complexities in creative ways. It further aids in a writer's ability to use logic and reason well.

To use images and drawings in the document, various desktop publishing tools such as Adobe Photoshop, Adobe Illustrator, screen capture tools, Paint, etc., can be used. In case technical writing demands graphs or statistical evaluation, it can be created using Microsoft Excel, Google Sheets, OpenOffice, Libre Office, Minitab, etc.

Medium of Publishing or Releasing

The method or medium of publishing is largely dependent on the types of products. Traditional choices include print or paper-based documents. Modern ways of publishing an article or a document include publishing e-books, PDF formats, web publishing, email, an organization's intranet, or mobile applications.

Releasing an article in a scientific journal could be one way to release an article for students, scientists, or by an organization. The writer or a team can explore relevant scientific journal publishers who provide a platform to release your content. You can find various platforms using search using search engines. Scientific journal publishers provide guidance for publishing content which includes but is not limited to writing templates, topics accepted by them, timelines, and other terms and conditions.

Releasing products useful for pharmaceutical companies, engineering companies, medical diagnosis, and chemical industries require promotion through high-quality technical content with facts and figures. The publishing of these contents could be done by an in-house marketing team or by choosing an appropriate media partner. The medium or platform for publishing the technical document can be emails to existing customers or the audience who are willing to receive mails on updated technologies or publishing the white paper on a company website.

Companies that are in the business of manufacturing and service publish their user guides on various platforms, such as company websites, mobile applications, leaflets to be provided with products in single or multiple languages.

Risk Register

Planning is an aid to the success of the project. It helps keep track of key milestones and timelines. Despite well-planned projects, there can be hiccups if potential risks are not anticipated during the planning phase. A risk register is one of the tools used for risk management. Risk management is a systematic process of proactively identifying potential risks, analyzing them, and taking mitigation steps to reduce or control the risk. This helps to stay on top of potential issues that can delay or skew the intended outcomes. The risk register includes all information about the nature of the risk, level of risk, what controls are in place, what could be the potential impact, and who owns it. The typical process flow risk management process is explained using Figure 5.

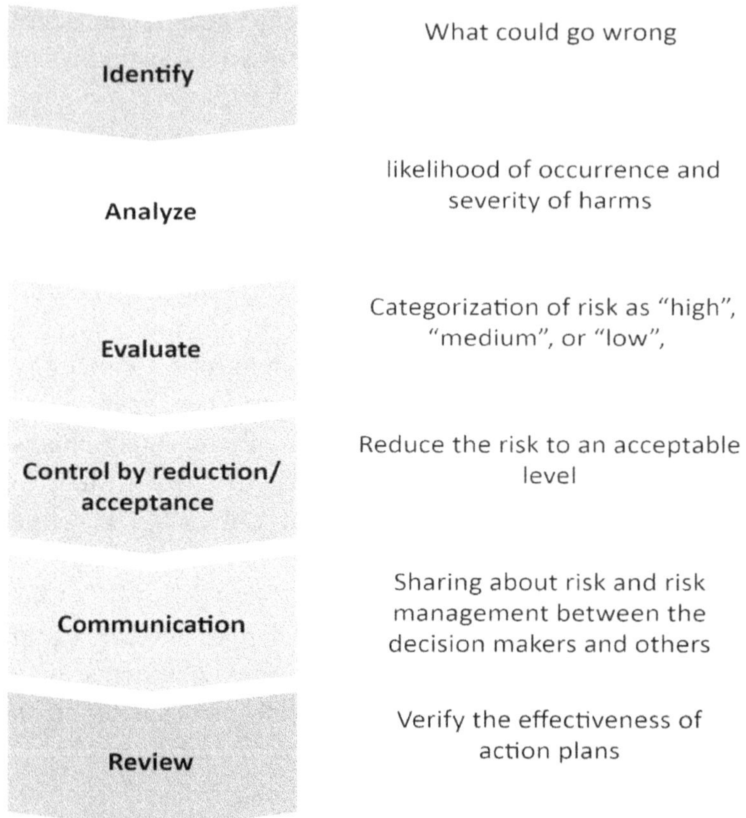

Figure 5: typical process flow risk management process

Using the above process, a risk register can be prepared. I have depicted an example of a risk register in **Figure 6**, which can be customized according to the process need. Apart from this, there are many risk management tools available, which are used for different aspects. Risk management in itself is an entirely vast subject, but here it is briefly discussed from the risk register preparation perspective as this is one of the important topics to have success in document writing.

Risk Register Scale				
4	4	8	12	16
3	3	6	9	12
2	2	4	6	8
1	1	2	3	4
	1	2	3	4

PROBABILITY (vertical axis) — IMPACT (horizontal axis)

Risk Register								
Anticipated Risk	Impact of Risk	Impact Level (1 to 4)	Probability Level (1 to 4)	Priority (Impact X Probability)	Mitigation Plan	Responsibility	Target Date	Open/ Close
More content and less time to achieve the target	The publication could not be completed resulting delay in product release. Market share of the product may decline	4	3	12	Identify the internal talent Or Engage the additional writer in advance	Human resource team Program manage to escalate Management to approve	July 25, 2020	Closed
Translator is not available	The publication could not be released in specific market resulting business loss	4	2	8	Engage the translator in advance with agreement or appoint a permanent position	Program manager to escalate Management to approve	August 25, 2020	Open

Figure 6: Example of Risk Register

While planning mitigation actions, a cost-benefit or effort-benefit analysis should be done to ensure that the mitigation plan is economical enough for the project viability.

Once planning is done considering all the applicable factors outlined in this chapter, you are good to go to the next step, the framework.

<table>
<tr>
<td></td>
<td>Clearly define the scope of the document to avoid potential failures or project drift. Use the 5W2H method to precisely outline what is in the scope and out of scope.

Use a mind map to organize content logically. It is a visual tool that helps in structuring the information effectively.

Involve Subject Matter Experts (SMEs) intermittently during the writing process for accurate content and practical insights, speeding up writing, and avoiding potential failures.

Clearly define roles and responsibilities at the beginning of a project to ensure a smooth workflow and timely completion of the document.

Develop a risk register to identify, analyze, and mitigate the potential risks during the planning phase, to avoid potential failure or delays.</td>
</tr>
<tr>
<td></td>
<td>*Summary:*

Think of this chapter as your map to project victory. It's where we join forces (stakeholders, unite!) to define scope and avoid nasty roadblocks. The 5W2H method is our trusty GPS, setting clear boundaries and keeping us on track. Once all set, dive into the info ocean, using mind maps as our magic tool to organize everything neatly.

Imagine writing about anatomy – would doctors and nurses need the same level of detail? Therefore, the scope really matters!</td>
</tr>
</table>

	And because even the smoothest journeys have bumps, a risk register acts as a treasure map for potential dangers. This risk assessment tool will help us to plan risk mitigation strategies.

Chapter 5: Framework

"A framework is like a good map; it shows you where you are and how to get where you want to go."

~Rita Mae Brown

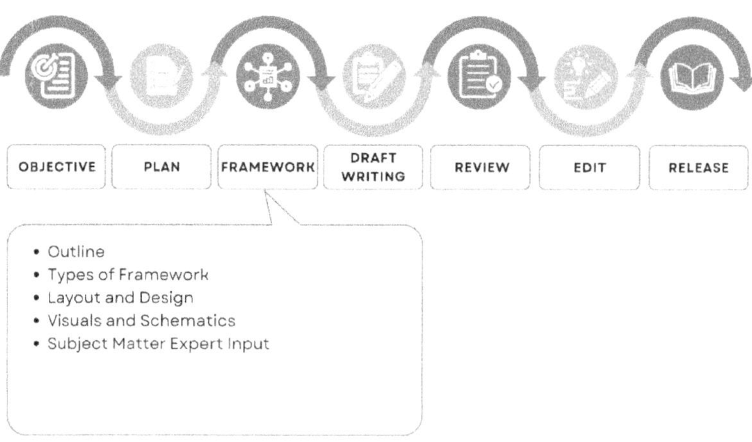

Once you've identified the purpose, the problem, the audience, and the type of document, it's time to gather your information. Crafting a framework is crucial to making your document coherent and easy for your readers to follow. A predictable and logical framework guides your readers through the information you provide in the document.

Outline headings and organize logically

To design your document's framework, start by creating an outline with headings and subheadings for each topic. During the initial planning phase, you've already done some mind mapping and drafted a table of contents for your technical document. Now, organize these topics logically to structure your document. One way to present your document's framework is by creating a table of contents, which will serve as a roadmap for your document.

However, a table of contents may not be necessary for all types of technical writing. For instance, if you're writing a Standard Operating Procedure (SOP) within an organization, a template with predefined headings and subheadings is likely already available. All you need to do is fill in the relevant information in the respective sections. Similarly, writing an article for a newsletter or a magazine wouldn't require a table of contents. Outlining your headings and subheadings logically will help your readers understand your message.

Types of frameworks

There are various methodologies you can use to frame your documents. In this book, I'll describe the following basic frameworks used for document writing:

- Descriptive
- Sequence
- Compare and Contrast
- Cause and Effect
- Problem and Solution

- Compendium
- User Guide

Depending on the document you're writing, you might use a single approach or a mix of approaches for document writing. The main goal of the framework is to organize the text in such a way that the reader can understand the content.

It's a good idea to interact with the people who could be the best fit for your book's audience. This is easy when the document is intended for internal use. When the document is intended for an external audience, you or a team member can put yourselves in the audience's shoes. During product development, the marketing and product development teams also research users' needs and wants. You can leverage the knowledge already acquired by the technical and marketing teams to develop your content and framework. As you progress, your table of contents will evolve and improve.

Descriptive Framework: The Descriptive framework is like to writing an essay or a story.

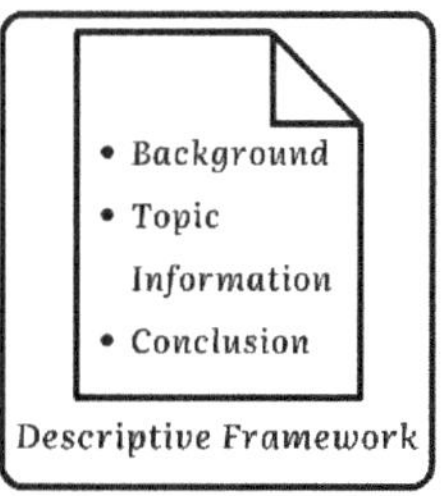

It begins with an introduction, followed by the body, and concludes with a closing statement. This framework involves explaining a topic or subject with background information, precise details, comprehensive explanations, attributes, and examples. For instance, you might start the document with a brief overview of the target topic, followed by supporting information, arguments, logic, examples, or case studies, and finally conclude with a summarizing statement.

Sequence Framework: The Sequence framework presents a series of process steps or events.

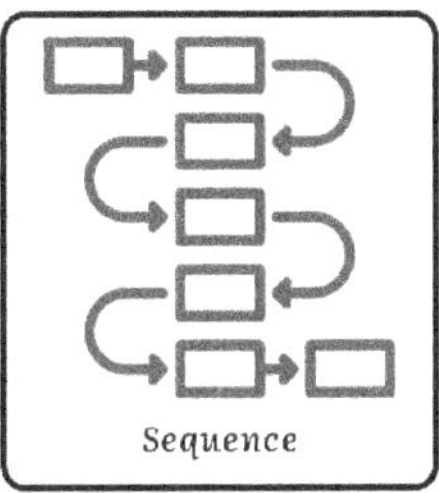

This framework is generally used for defining the standard operating procedure of equipment or outlining a process. It can also be used to describe a part of an event investigation where a chronology of events needs to be drafted.

Compare and Contrast Framework: The Compare and Contrast framework is used when you want to discuss the similarities and differences between two or more objects, events, or places.

Here are a few examples of where this framework can be used in technical writing:

- To describe the advantages of a new version of a product compared to the older version

- To compare the product with competitors' products and explain the benefits of the product you are writing about

- While writing scientifically, to discuss the pros and cons of different theories, techniques, or methods

Cause and Effect Framework: The Cause and Effect framework is useful when an effect (a fact or event) and its cause (something responsible for the fact or event) need to be documented.

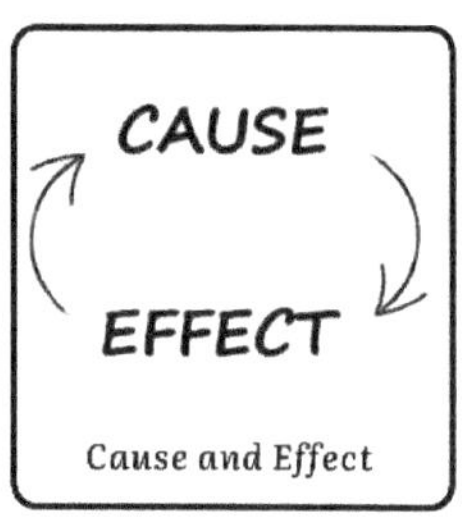

Cause and Effect

This framework is useful for documenting various scientific facts, conducting root cause analysis during event investigations, presenting research outcomes, writing theses, explaining mechanisms of actions, and other technical documents. It's important to note that one cause can have multiple effects or multiple causes could yield a similar effect.

It's not always possible to establish a relation between cause and effect. In such cases, the conclusion can be derived based on available information, and the cause is documented as a potential cause or the most probable cause.

Problem and Solution Framework: The Problem-Solution framework is incredibly useful in technical and professional documents.

A great starting point for any document is to define the problem and solution perspective. Here are a few examples where the problem-solution approach can be applied in technical writing:

- Designing a Standard Operating Procedure for machine operation with troubleshooting

- Designing a troubleshooting guide for a product

- Designing a brochure that defines a problem and provides a solution through your product

- Publishing a research article that identifies a gap in current knowledge and proposes possible solutions

In this framework, it's crucial for the writer to carefully organize the draft. Explain to your readers why the problem is important and needs to be solved. Clearly present the solutions so that your reader is convinced. The solution you offer should be practical, feasible, and easy to implement. It should also be cost-effective and time-efficient. Most importantly, it's essential to convince your readers that your solution is the best compared to others.

Compendium Framework: A compendium, as defined by the dictionary, is "a collection of concise but detailed information about a particular subject, especially in a book or other publication".

This framework is ideal for publishing documents such as scientific journals, technical newsletters, technical magazines, blogs, and other similar documents that contain distinct topics within one document.

In this framework, a document may contain entirely different topics or topics that are bucketed into different categories. For example, a magazine that publishes upcoming technologies of all kinds. In such a magazine, each topic is unique and standalone, but it can be bucketed into categories like electronics, electrical, automobile, construction, and so on.

User Guide Framework: The User Guide framework is useful for preparing technical documents such as manuals or user guides for a product.

These types of documents are generally required when complete product information needs to be provided with the product. For example, a pharmaceutical company procuring equipment to produce tablets. For such a complex equipment, equipment manufacturer provides complete information about the equipment. The document covers the standard operating procedure, equipment general arrangement layout, electrical diagrams, piping and instrumentation diagram, component list and its specifications, material of construction, general cleaning procedure, general maintenance procedure, troubleshooting, safety measures, and other relevant information.

Layout and Design

The secret to any good design lies in the arrangement and placement of text and visuals. If the layout and design of a document are not correctly understood, there's likelihood that the intended message will be lost. To design a good layout, the arrangement of visual elements and their relationship with the text is a crucial aspect. This is precisely what layout design is all about. The layout and design of text and visuals give meaning to your write-up and make it appealing. It helps maintain balance from page to page for the readers.

Visuals and Schematics

To communicate technical documents precisely, more than just text is required. Visuals and schematics are pictorial illustrations of information. Pictures, schematics, graphs, charts, and tables are some examples of visuals that can be used in technical documents. Text with visual aid will help readers understand the information quickly, accurately, and with minimal text. These visuals help supplement your written ideas and simplify complex text descriptions. Visuals engage readers, allow them to visualize the idea, and motivate them. Visuals are used to complement text in technical documents. The key point to remember is that visuals provide clarity, illustrate, and support your written text, but they are not a replacement for written text.

Visual elements are referred to as either Tables or Figures. A table is used to organize data in rows and columns. To present multidimensional data, a table is a good tool to use. It simplifies and reduces the text. It also helps compare data with each other and show the relationship between data, which could be single, dimensional, or multidimensional. Figures include charts, graphs, engineering layouts, photographs, and drawings.

Subject Matter Expert Input

Now you might be thinking, why is this heading here? Stay with me as we unravel the unexpected nuances of this chapter together. It is very important that intermittent review of the structure, table of content, and the approach chosen by the writer should be reviewed by subject matter experts as well as by stakeholders. This will provide assurance that the writer is moving in the right direction as per expectation.

The feedback from subject matter experts and stakeholders will refine the structure and table of content. If the situation permits, the structure could be run through the target audience or those who can give similar feedback.

The above review need not be a formal review but can provide important insight and confidence to the writer that the project is going in the right direction. This could save an enormous amount of time by avoiding rework at a later stage.

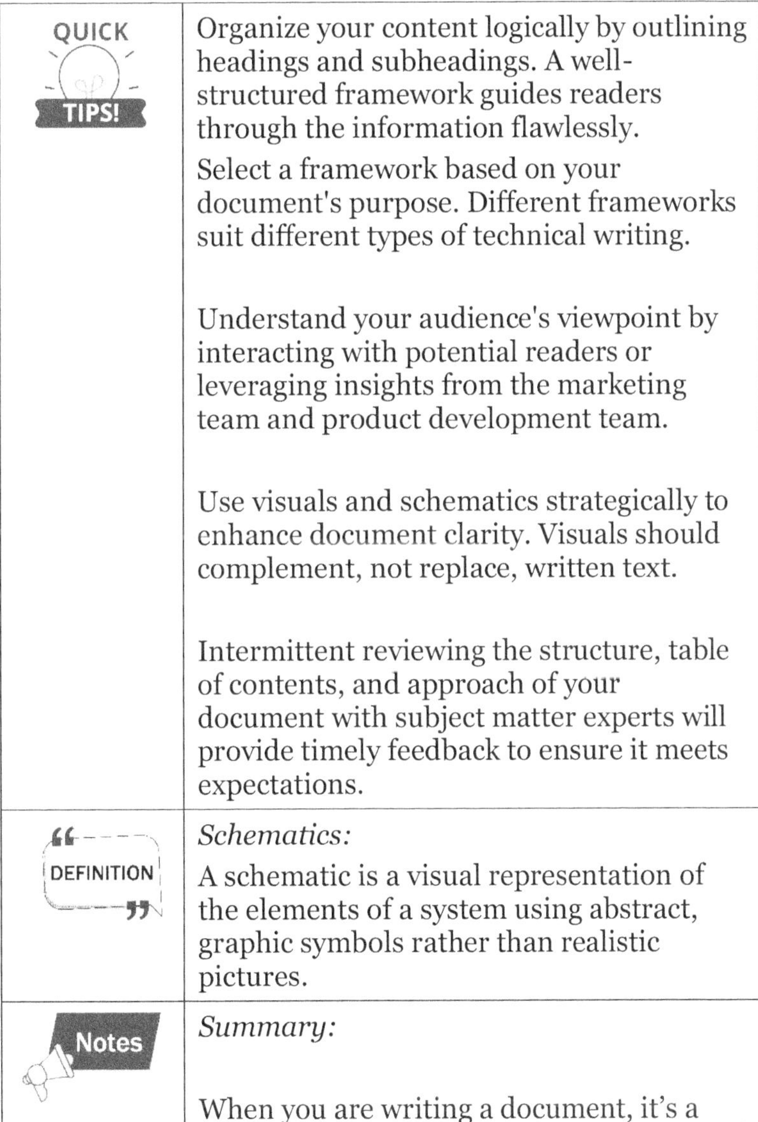

QUICK TIPS!	Organize your content logically by outlining headings and subheadings. A well-structured framework guides readers through the information flawlessly.
	Select a framework based on your document's purpose. Different frameworks suit different types of technical writing.
	Understand your audience's viewpoint by interacting with potential readers or leveraging insights from the marketing team and product development team.
	Use visuals and schematics strategically to enhance document clarity. Visuals should complement, not replace, written text.
	Intermittent reviewing the structure, table of contents, and approach of your document with subject matter experts will provide timely feedback to ensure it meets expectations.
DEFINITION	*Schematics:* A schematic is a visual representation of the elements of a system using abstract, graphic symbols rather than realistic pictures.
Notes	*Summary:* When you are writing a document, it's a

	journey. In this journey, the framework will act as the compass guiding readers. The different frameworks illustrated in the book will act as different navigational tools: descriptive, sequence, problem-solution, and more. Each framework works best in certain situations. For example, a user guide framework meticulously maps equipment manuals, while cause-and-effect illuminates scientific papers. Visuals act as helpful signposts. For example, tables organize data easy to understand form, and figures visualize complex concepts. Don't forget to refine your path! Collaborate with experts and potential readers to ensure your chosen path caters to your audience. Remember that the right framework makes your document from a confusing maze into clear and engaging content. Choose the framework wisely, and your readers will reach their destination informed and satisfied.

Chapter 6: Draft Writing

*"Almost all good writing begins with terrible first efforts.
You need to start somewhere."*

~ Anne Lamott

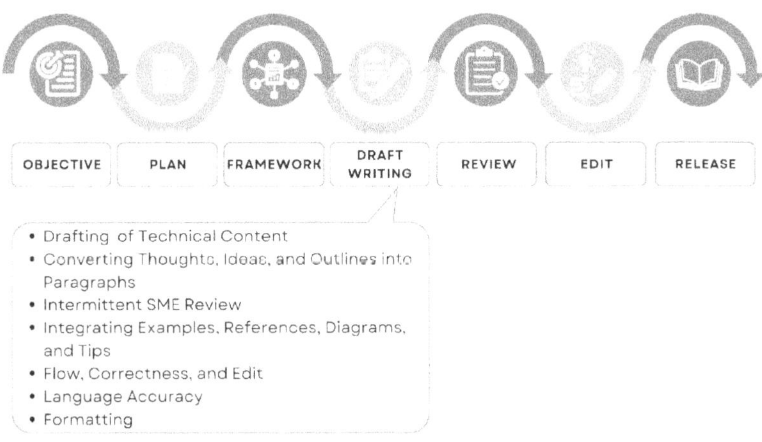

Once the groundwork is done, the drafting of the document commences. Draft writing is a cyclical process, undergoing multiple iterations until a polished draft emerges. The completion of the first draft signifies only half the work done. The journey to the publishing stage involves numerous reviews, edits, refinements, and finally, publishing post-approvals.

As you scrutinize technical literature, published books, or even within a company, you'll notice that documents are often the product of more than one writer. A collaborative approach proves more productive. Each writer can assume a different role such as author, editor, reviewer, and Subject Matter Expert (SME). Collaborative writing minimizes multiple revisions of drafts as the team can leverage each other's expertise, engage in discussions, and arrive at common conclusions.

Once the draft content is populated, it's time for a sanity check. At this juncture, ensure the flow of writing, language accuracy, and finally, formatting to make the document presentable.

Drafting and Concentrating on Technical Content

Drafting the first manuscript can be a challenging task. However, if you've done your homework thoroughly, it becomes significantly easier. You're well-prepared to write the first draft when you've established the objective, understood the audience's needs, planned for writing, completed research, and are ready with a framework with a draft table of content.

When you commence writing your first draft, it's not necessary to follow a sequential order from introduction to the final chapter.

Converting Thoughts, Ideas, and Outlines into Paragraphs

Simply transcribe and elaborate the notes from your outline, notations recorded during meetings, and SMEs' views into paragraphs, without worrying about spelling, grammar, or language. The focus should be on drafting the technical content first.

A few authors have developed different methods of writing that help to write faster, clearer, and make you a more productive writer. These methods can be a valuable addition to your writing process.

<u>Free writing:</u> In a book, "Accidental Genius", Mark Levy suggests the method of free writing. *"(Freewriting) pushes the brain to think longer, deeper, and more unconventionally than it normally would. By giving yourself a handful of liberating freewriting rules to follow, you back your mind into a corner where it can't help but come up with new thoughts. You could call freewriting a form of forced creativity"*.

This process emphasizes ideas and content over structure. The idea is to note down thoughts, refine them later to present in more meaningful ways, build upon them, and finally finalize the content with edits and polishing.

<u>The carpenter's method:</u> Jack Hart suggests Carpenter's method in his book, "A Writer's Coach". He quotes one of his workshop participants that *"Years ago I used to futz with every sentence, but then an editor told me something that really made sense. He said that when a carpenter builds a piece of furniture he doesn't first make one side, perfect that, and then construct another side and perfect that. He must build the entire frame and then go back and put the finishing touches on each section. Even when I am on deadline, I think of what I write first as an imperfect frame that will be improved later"*.

The theme behind the above story is that you don't need to edit and polish each statement you write instantly. Allow ideas to flow in the mind, outline it, draft it, take a breath and revisit it and finally edit to make the content ready for final draft.

The Knitter's method: The knitter's method is different from above both the method and opposite. As per this philosophy, make each part of the content perfect before going ahead.

As Mark McGuinness writes, *"One of the sacred cows of the creative thinking industry is that we should separate idea generation, execution, and evaluation, so that they don't interfere with each other. But my experience as a writer and coach suggests that this isn't how many creative professionals work.*

When I'm writing, I'm reading, evaluating, and tweaking as I go. I'll write a few sentences then pause and go back to read them through. Sometimes it's immediately obvious I haven't quite captured the thought or image, so I'll make a few changes before I go on. If I get stuck, I'll stop and read through the whole piece, trying to pick up the thread of inspiration where I lost it. Once I see where I got tangled up, it's a relief to untangle it and get going again."

Advanced writers could use the Knitting Method efficiently. However, beginners and semi-advanced level writers carry a risk of spending a lot of time and potentially missing project deadlines.

Intermittent SME Review or Technical Resource Review

Upon completing the first draft, it's time to seek feedback from Subject Matter Experts (SMEs) and accessible technical experts. The interim review aims to evaluate the work done so far and gather recommendations to enhance the technical content in terms of accuracy, flow, and practicability. This independent advice will provide insights about the content's relevance in line with the subject, topic, and whether the document is technically and scientifically correct.

The review could take the form of a mock trial of the document with the assistance of technical/ product experts, especially if the writing pertains to user equipment procedures, equipment testing or qualification protocols, user guides, etc. The review could be walkthrough-based, sending documents to someone who fits into the user profile, or a peer review.

Reviewing is a time-consuming process. If someone is an SME or a technical expert, securing their time for review requires effort. Each reviewer has their own style and prerequisites to start the review. Some prefer to have the full document or full topic at once, while others may review the document in parts and provide feedback. To facilitate your task, it's crucial to understand the individual SME or technical expert's review style and involve them accordingly.

There are always ways to get your work done, and one of the best ways is to help them help you. If the SME or technical expert has some work that you could do, you can always offer your help so they can spare some time to support you in the review process.

The primary objective of the intermittent review is to receive feedback at the right time, so you're not caught off guard at the end, necessitating a rework from scratch.

Integrating Examples, References, Diagrams, and Tips

Once your technical content is ready, enhance its usability with the help of examples. Support your arguments with references and literature. Provide tips that will offer topic-specific insights and serve as a quick guide for someone revisiting the document.

In the case of scientific documents, an example can help the reader correlate the text with a real-life scenario, thereby facilitating a better understanding of the concept. As the saying goes, "An example is worth a thousand words." When you read any law of physics for the first time, it might be hard to understand, but when you read an example, it becomes easier.

Referencing is a crucial part of technical work. The breadth and depth of your work can be gauged based on your research, and it acknowledges others' work. You should give credit whenever you use someone else's work. Giving references is a best practice in academic writing. The key benefits of referencing include:

- Strengthens the topic and supports your arguments and views

- Demonstrates your research on the topic

- Helps to avoid plagiarism

Assume you have two books: one with visuals and diagrams and another with only text. Which one would you prefer to pick first? The obvious answer for most of us would be the one with visuals.

When it comes to technical writing, it is expected that the content should be supported with visuals to facilitate understanding. The use of visuals should be optimal to maintain a balance between content and supporting graphics. The correct use of visuals has many advantages:

- Helps readers understand complex information quickly and easily

- Immediately attracts the attention of readers

- Makes it easier to remember and helps retain information for a longer time

- Saves the reader's time

The important characteristics of visuals are as follows:

- **Self-explanatory:** Visuals should be self-explanatory so that they do not require text to understand. Providing legends, names of components in the visual, ensuring all parts of visuals are clearly visible and legible, and providing a proper title.

- **Linkage with the text:** When referring to the visual, the terminology used to explain the content should align with the titles and other nomenclature used in the visuals. It is always a preferred practice to provide numbering to the visuals and link that with the text. For example, Diagram - 1, Photograph - 5, Drawing - 12, etc.

- **Location of visuals in the content:** Locate the visuals in the content at meaningful places. The reader should not have to change the pages frequently while reading the text and simultaneously referring to the visuals. The size of the visual is equally important so that it fits in the text in such a way that it does not create discomfort while reading.

- **Effortless:** When dealing with technical writing, it does not require fancy fonts or fancy visuals. Keep it simple with only the required information. When visuals are filled with too many things, it's hard to understand and becomes more confusing rather than helpful to the readers. Do not create your own visuals if not necessary. Visuals can be taken from reference material. Do not forget to indicate the reference from where it is used.

Providing short notes, tips, summaries, and insights at the end of each topic or at important content will be very helpful for the readers. This will summarize the content in a few words, and help the reader to remember. When the reader wants to refresh the content after a while, it will be very helpful and within no time, the reader could re-memorize it.

Review for Flow, Correctness, and Edit

Once your draft content is ready, take a pause. Allow yourself some time before reviewing it with the audience in mind. This change in perspective may reveal plenty of room for improvement.

Go through the document line by line and check for the flow of writing. Verify whether it adheres to the expected structure you had in mind. You may find that the table of contents requires some changes based on feedback from SMEs and technical experts.

Check for necessary corrections on technical aspects and other requirements as per your documentation plan. Ensure the completeness of the required content. Writing and reviewing is a cyclical process until you reach the final draft. After the review, when you find that enough work has been done and nothing more needs to be added, you are ready to proceed to the final steps of draft writing, which include ensuring language accuracy and formatting the document.

It's important to accept that whatever is written will never be perfect. Focus on task completion with all its aspects and strive to do the best you can within the given time. Getting caught in a loop of write-review-write with the aim of achieving perfection can lead to never-ending writing and the task may never be completed within time.

Language Accuracy

As a technical writer, it's crucial to consider not only what you express but also how you express it. Effective communication requires more than just providing accurate and comprehensive content; one must also consider language criteria and the ease with which readers can understand the document.

According to the document Federal Plain Language Guidelines, March 2011, Rev. 1, May 2011, *"One of the most popular plain language myths is that you have to "dumb down" your content so that everyone everywhere can read it. That's not true. The first rule of plain language is: write for your audience. Use language your audience knows and feels comfortable with. Take your audience's current level of knowledge into account. Don't write for an 8th grade class if your audience is composed of PhD candidates, small business owners, working parents, or immigrants. Only write for 8th graders if your audience is, in fact, an 8th grade class"*.

Using plain English is key to clear and concise writing. The simplicity of text is an important prerequisite for the readability of a document. A technical writer should always strive to write in a plain, clear, and direct language to make it clear who's doing what. Use plain words so that the vocabulary is easily understood. Use good standard English by using accurate grammar and punctuation.

Clear writing is evidence of clear thinking. Adhering to these guidelines will help ensure clear and concise writing.

In this section of the chapter, we delve into the fundamentals of English grammar. Through illustrative examples, I aim to simplify the complexities of technical writing. My goal is to help you minimize errors and enhance your reader's experience. Let's embark on this journey to make your writing more effective and engaging.

Sentences: Choose words carefully while writing. Main idea should come first. Place your words carefully. Word order does matter significantly.

- **Write short sentences:** Keep statements short and simple. A sentence should contain a single piece of information. Avoid compound sentences. Long, complicated sentences are often misunderstood and it seems you are not clear what you want to say. Breakdown the statements; shorter sentences are better to communicate complex information.

  Too many dependent clauses confuse the audience by losing the main point between many words. Don't try to put everything in one sentence. It will lose the sense of idea.

  Simple sentence: He worked hard to clear the interview.

  Complex sentence: He worked hard so that he might clear the interview.

- **Keep subject, verb, and object (SVO) close together:** *"The communicative strategy found in the SVO word order can be considered listener-oriented because the speaker or writer, who has new* *information to communicate, considers more important the fact that the message is clear to the hearer than his/her necessity to communicate (Siewierska, 1996: 374)."* (Maria Martinez Lirola, Main Processes of Thematization and Postponement in English. Peter Lang AG, 2009)

Naturally speaking, the order of an English sentence is subject-verb-object. This is how we first learned to write sentences. When we use modifiers, phrases, or clauses between the essential parts of a statement, it becomes confusing for the user to understand you.

Avoid double negatives and exceptions to exceptions: We are familiar with speaking positive statements. Double negatives mean two negative words in the same sentence. Using two negatives usually turns the thought into a positive one. Such statements are usually discouraged in English. Such statements are considered to be poor grammar and they can be confusing.

Example of double negatives: The outcome is not inconclusive.

Example of positive statement: The outcome is conclusive.

Here are a few more examples that will help you understand how to avoid making complex statements.

Double negative → Positive
no fewer than → at least
has not yet attained → is under
may not ...until → may only ... when
is not ... unless → is ... only if

Similar to the above, an exception that contains an exception also forms double negative. Such statements are harder for the user to understand. It would be prudent to make the statement positive to make it reader friendly.

- **Placement of main idea, exceptions and conditions in the statement:** Statement becomes easier to read when it starts with the main idea. When an introductory phrase begins with "except," it becomes harder to grasp and the reader is required to re-read the statement to establish correlations. Write-up becomes easy to read when it starts with the main idea first and then exceptions or conditions.

  There is no rule on where to put exceptions and conditions, however, it should be placed in such a way that it becomes easy for readers. If an exception or condition is concise and its early placement in the statement can prevent confusion, it should be placed at the beginning rather than the end.

  If an exception or condition is a long statement and the main idea is short, put the main clause first and then state the exception or condition.

  If there are numerous exceptions or conditions, they can be listed using bullet points or numbers below the main statement.

- **Place words carefully:** The order of words in an English sentence is very important, because a change in word order may result in a change of meaning or fluency.

  Generally, sentences have a subject, and then something that is said about the subject, which is usually the rest of the sentence. This divides the sentence into the subject-verb-object.

Inappropriate word placement can cause confusion and ambiguity. For example, use modifiers next to the words they modify. In the following statement, "only" is used as a modifier and describes about "following".

Incorrect statement: You are only required to provide the following.

Correct statement: You are required to provide only the following.

Paragraphs: Similar to the longer statement, long paragraphs are not considered as good writing practice. A paragraph should contain a single topic or idea. Length of the paragraphs should be about 200 words. On the other hand, it should not be very short.

- **Have a topic sentence:** The heading of a paragraph helps to provide context to the reader about what they are going to read. After providing this context, detailed information should follow. If details are provided to the reader first, it is likely that they will not read to the end and may miss important information.

  Busy audience will skim the document and will read details only that they want. A good introduction of paragraph will help readers to find what they want. The document should be designed in such a way that the reader should be able to get good general understanding by skimming your topic sentences.

- **Use transition words:** Whether it is general writing and technical writing, the aim is to convey information clearly and reader to follow your way of thinking. In other words, transitions tell readers how to follow the information you presented to them. Transitions word establishes logical connections between sentences, paragraphs, and sections of your document. Transition word or phrase clearly tells the audience whether the paragraph linked to the previous paragraph, stating new one in sequence, contrasts with previously provided information, or takes the topic to a completely different direction.

- Bryan Garner (2001) divides transition words into three types:

  i. Pointing words refer directly to something already mentioned. Examples of pointing words are "this", "that", "these", "those", and "the".

  ii. Echo links and words or phrases echo a previously mentioned idea.

  iii. Explicit connectives are words whose main purpose is to supply transitions. Examples are "such as" "further", "also", "therefore".

Following are types of transition and its expressions:

Opposition/ Contradiction/ Exception: but, however, in spite of, despite, on the other hand, while, nevertheless, nonetheless, notwithstanding, in contrast, on the contrary, still, yet, conversely, instead, besides

Similarity/ Agreement/ Addition: also, in the same way, just as, likewise, correspondingly, similarly, additionally, equally

Example: in other words, for example, for instance, namely, specifically, to illustrate, such as, frequently, including

Place/ location: here, above, in front of, adjacent, below, beyond, here, in front, in back, nearby, there, beneath, across, adjacent to, opposite to, around

Emphasis: even, indeed, in fact, of course, truly, notably, must be remembered

Cause/ Effect/ Condition: accordingly, in effect, under those circumstances, consequently, hence, so, therefore, thus, in that case, therefore, accordingly

Additional Support or Evidence: additionally, again, what is more, similarly, further, also, and, as well, besides, equally important, further, furthermore, in addition, moreover, then

Chronology/ Sequence/ Order/ time: first, second, third, next, then, finally, after, afterward, at last, before, since, currently, during, earlier, immediately, until, later, meanwhile, now, recently, simultaneously, subsequently, then, once, suddenly

Conclusion/ Summary: to summarize, in summary, in a word, in brief, briefly, in conclusion, in the end, in the final analysis, on the whole, thus, to conclude, to summarize, in sum, to sum up, finally, to sum up, to conclude, in conclusion, in short

Use active voice: Active voice is a natural language which we speak and easy to understand by us. Active voice provides clarity about who needs to do what and eliminates ambiguity regarding responsibilities. Passive should be used when action is more important than who is doing the action.

When one is writing Standard Operating Procedures (SOPs) or user guide, using active voice is preferable. In an active sentence, the person acting is the subject of the sentence. However, it is not important when describing anything such as scientific fact, background of a document, or where the subject is not important.

Use the simplest form of a verb: The simplest and easiest form of a verb is present tense. A document written in the present tense makes it more direct and helps to make your point clear. The simple present is used when the precise beginning or ending of the action, event or condition is unknown or not important to the meaning of the statements.

Sometimes, you may need to use other tenses. For example, describing the history and future prospective of a situation, thing or act.

Punctuation: Punctuation plays a significant role in the meaning statements. Misplacement of punctuation will change the meaning of the sentence. Everyone finds a challenge in placement of punctuation, at least occasionally. When you are not sure, you can always use handbooks, as well as a variety of websites.

- **Spacing requirement with punctuations:** Rule for spacing after punctuation is one space followed by commas, periods, semicolons, colons, exclamation points, question marks, and quotation marks. When using a hyphen, no space either side of hyphens. A detail about hyphen is discussed in the later in this section.

Examples:

He is Dr. John. However, he does not practice medicine.

In his career he worked as a professor, consultant, writer etc.

- **Periods:** A period is used in a sentence after a complete statement. If the last word in the sentence ends with a period, another period is not required. While asking an indirect question, a period is used instead of a question mark.

Examples:

He asked where his laptop was.

His laptop was with me.

- **Ellipsis marks:** Use an ellipsis when removing a phrase, word, clauses, line, paragraph, or more from a quoted statement. Ellipses remove words that are less relevant or not required to specify. They are useful in discussing to the point without distraction from a quoted paragraph. An ellipsis is a punctuation mark with three dots. Use three dots for omission in the middle of a statement or between statements.
The punctuation can also be used to express hesitation, changes of mood, or opinions trailing off.

Examples:

Without ellipsis: Today, after the board meeting, we promoted him as a head of department.

With ellipsis: Today ... we promoted him as a head of department.

- **Commas:** Out of all punctuation marks in English, this punctuation is most confusing and misused. This happens because there are many rules around it. A comma indicates a pause within a sentence.

<u>Commas between subjects and verbs:</u> Commas should not be inserted between the subject of a sentence and the verb. A comma indicates a break in the sentence and since the subject is linked to the verb (the subject carries out the action described by the verb) it is incorrect to have a break between these two elements.

Examples:

Incorrect: My friend Ray, is a wonderful writer.

Correct: My friend Ray is a wonderful writer

Incorrect: The work that gives me money, may also affect my health.

Correct: The work that gives me money may also affect my health.

<u>No comma between the two nouns, noun phrases, or noun clauses in a compound subject or compound object</u>

Examples:

Incorrect: Rose, and her friend will be going at river front next Sunday.

Correct: Rose and her friend will be going at river front next Sunday.

<u>No comma between the two verbs or verb phrases in a compound predicate</u>

Examples:

Incorrect: I wanted to go to movie show, but ran out of money.

Correct: I wanted to go to movie show but ran out of money.

<u>Use commas for independent clauses with the help of coordinating conjunctions:</u> When you want to join two independent clauses, you can use coordinating conjunctions. Examples of coordinating conjunctions are so, yet, and, but, for, or, nor.

Examples:

Soni tried a new diet, but she still gained weight.

The interviewee explained his question, yet the employer still didn't seem to be convinced.

<u>Commas after introductory elements:</u> An introductory clause or phrase acts as qualifying or background information about the main sentence. Introductory clauses start with adverbs such as before, when, while, after, as, although, because, if, since, though, until, etc.

Examples:

If the team wants to win, all players must practice every day.

Similar to introductory clauses, introductory phrases also set the context for the main action of the sentence, but they are not complete clauses. Phrase a group of words that adds meaning to the statements and do not have a subject and a verb. Clause is a part of statement that contains a set of words having a subject and a verb.

Examples:

If the team wants to win, all players must practice every day.

To stay in shape for competition, all players must practice every day.

Similar to introductory words and clauses, introductory words require a comma. Common introductory words are adverbs, interjections, and sometimes names. Common introductory adverbs are however, meanwhile, suddenly, finally, besides, and still. Commonly used interjections are please, thanks, yes, no, well etc.

Examples:

Incorrect: Meanwhile he read two books of his favourite author.

Correct: Meanwhile, he read two books of his favourite author.

Incorrect: Let's eat friends!

Correct: Let's eat, friends!

Incorrect: Yes I will, thanks.

Correct: Yes, I will, thanks.

<u>Comma while comparison:</u> Comma is not required to add when you are doing comparison using "than"

Incorrect: This table is lighter, than that cupboard.

Correct: This table is lighter than that cupboard.

<u>Comma with a question tag:</u> A question tag is a short phrase added at the end of a statement.

Example: The winning team looks very happy, aren't they?

<u>Comma to direct address:</u> Comma is used when addressing any person by name.

Example: Kent, there's someone at the door for you.

<u>Commas in Dates:</u> In certain formats of dates, we use commas.

Example: October 25, 2020

<u>Commas to separate two or more coordinate adjectives:</u> Coordinate adjectives are adjectives which describe nouns with equal degree.

Example: That person is an arrogant, self-righteous, irritating idiot.

<u>Comma with but:</u> Comma before the word but is used when it is joining two independent clauses.

Example: Roy is a good writer, but he's an even better consultant.

<u>Comma with and</u>: When you have a list that contains more than two items, use a comma before the and. Comma is also used before and when joining two Independent clauses.

Examples:

Incorrect: My friend is smart, and simple.

Correct: My friend is smart and simple.

Correct: She loves ice cricket, chess, and table tennis.

Correct: John took math lessons for sixteen months, and today he is an expert.

In the case of listing more than two items, the final comma that comes before the 'and' is optional. This depends on the style the writer uses.

<u>Comma in statement with non-restrictive Clause:</u> A non-restrictive clause is a clause that provides additional information which is non-essential information.

Example:

Vickey, who is my very good friend, is a brilliant doctor.

<u>Comma in statement with restrictive Clause:</u> Restrictive clauses are generally used with who or that. Comma is not required to be used in that case.

Example:

Incorrect: The book, that teacher suggested, is a very helpful resource for the course.

Correct: The book that teacher suggested is a very helpful resource for the course.

<u>Comma with quotation marks:</u> With American English, a comma comes before the closing quotation mark. However, in British English, it is after the quotation marks.

Example:

American English: "Give me a book," said Rosan.
British English: "Give me a book", said Rosan.

<u>Comma is not required for following</u>:
No comma between an article and a noun.

No comma with "as well as" when it is part of a non-restrictive clause.

No comma with "such as" when uses as a restrictive clause.

Comma before "too" is optional.

No comma requires between the verbs or verb phrases in a compound predicate.

No comma to separate the subject from the verb.

- **Semicolons:** Semicolon is used to connect related independent clauses.

Example: Methyu has gone to the gym; Sweety has gone for swimming.

Also, a semicolon links up by replacing a conjunction between two related clauses.

Example:

I saw a dangerous lion, and it was trying to catch a rabbit.

I saw a dangerous lion; it was trying to catch a rabbit.

Use a semicolon to divide items in a list if the items contain internal commas.

Example:

Last year they have done business tour to Delhi, Tokyo; Japan, United Kingdom; and Toronto, Canada.

- **Colons:** Colons are used for independent clauses when second explains or supports the first

Example: I have very little time to complete the course: my ticket for London is in one month.

Colons used for salutations and isolated elements

Example:

Followings are different elements where colons are used to separate the details.

Ratios:
50:50

Time:
10:20 p.m.

Chapters and verses:
2:2-1

Titles:
Spark Notes: To Kill a Mockingbird

Mail openings or after greeting in letter:
Dear Ms. Fernandes:

Introducing a list:
The teacher specializes in many subjects: Pharmacology, Pharmaceutics, and Pharmaceutical Chemistry.

- **Question marks:** The question mark is used at the end of a direct question. For indirect questions, period is used.

Examples:

Direct question: What is he doing studying?

Indirect question: I wonder what he's studying in college.

In contrast, requests or commands that are phrased as questions are end with a period. These are not true questions.

Example:

Would you please send this letter to the head office.

- **Exclamation marks:** An exclamation mark is used to express surprising fact, forceful or to show emphasis.

  Exclamation marks are generally not used in formal letters.

  Example:

  Our team won the final match!

  An exclamation mark is used as an interjection. An interjection used as part of a sentence that expresses sudden emotion such as surprise, joy or enthusiasm.

  Examples:

  Oh!
  Stop!

  Help!

  Wow!

- **Parentheses:** Round brackets are also called parentheses. This is mainly used to separate information that is not essential to the meaning of the sentence, but, it provides additional information or clarity to the parent statement. The statement without bracketed information in the sentence would still make perfectly sense. If information in parentheses ends a sentence, the period goes after the parentheses.

Examples:

I went to Italy (The capital of Italy) last year.

He has many bikes (10).

- **Apostrophes:** Apostrophe is used to show possession and for contractions. For the possession, place the apostrophe before the s to show singular possession. To show plural possession, make the noun plural first and then immediately use the apostrophe. In case of names ending with s or an s sound are not required to have the second s added in possessive form.

Examples:

Contractions: doesn't, isn't

Singular possession: This is Indila's car.

Plural possession: Three boys' pen

No apostrophe for the plural of a name.

Examples:

The Johns have two books and a pen.

No apostrophe with possessive pronouns as they already show possession.

Examples: such as his, hers, its, theirs, ours, yours, whose.

Correct: This bicycle is hers, not yours.

Incorrect: Sincerely your's.

- **Hyphens:** A hyphen is a punctuation mark that is used to join words. This is called a compound adjective with a single meaning idea. If you are not sure, use a dictionary.

  Examples:

  Well-respected teacher

  State-of-the-art facility

  When numbers are used as the initial part of a compound adjective, hyphen is used to connect them.

  Examples:

  Principal gave a 10-minute speech to the children

  He stays at the four-story building.

  Hyphen is also used with Prefixes, such as Ex-, Self- etc.

  Example: I had a lunch with my ex-boss!

- **Dashes:** A dash is a horizontal line; little longer than a hyphen. It is generally used to indicate a range or a pause. There are two forms of dashes: em (—) and en (–). No spaces required before or after en or em dashes.

  En –) is used for periods of time.

Examples:

The years 2019–2021

February–June

An en dash is also used in place of a hyphen when combining open compounds.

Examples: USA–Canada border

In informal writing, em dashes may replace punctuations such as semicolons, commas, colons, and parentheses to indicate added emphasis.

Examples:

You are the friend—the only friend—who offered to help me.

Never have I met such a brilliant scientist—before you.

- **Capitalization:** Following are rules of capitalization.

  o Capitalize the first word in a sentence.

  o Capitalize the pronoun "I".

  o Capitalize proper nouns. Those are places, organizations, people, and sometimes things.

     Examples: Mumbai, Pepsi, John, Honda City

- o Capitalize family relationships when used as proper nouns.

  Examples: Uncle Peter, Grandma Rose

- o Capitalize titles that appear before names, but not after names.

  Examples: Chairperson July

- o Capitalize the person's title when it follows the name on the address or signature.

  Example:

  Sincerely,

  Mr. Joy, Vice President

- o Capitalize the titles of high-ranking government officials when used with or before their names.

  Attorney General Henry

- o Capitalize directions

  Examples: North, South, East, and West when used as sections of the country.

- o Capitalize the days of the week, the months of the year, and the holidays

- o Capitalize members of national, political, racial, social, civic, and athletic groups.

- o Capitalize periods and events, but not century numbers.

So that would be the "Greek Era"

- o Capitalize trademarks.

  Examples: Toyota, Microsoft

- o Capitalize any title when used as a direct address.

  Will you give me medicine, Nurse?

- o Capitalize the names of specific courses.

  Jolly English

Use "must" to indicate requirements: When you want to emphasize the mandatory requirement, use "must". This will indicate that to action or requirement is mandatory to do.

Use pronouns to speak directly to readers: The document that you write may be applicable to all readers. However, you are communicating with the one person who is reading the document. The use of pronouns helps psychologically connect the reader with the document. Using 'you' has a greater impact on readers, drawing them into the document and making it more relevant to them. In the document, 'you' reflects responsibility. Using pronouns makes sentences simpler and easier to read and understand.

<u>Don't say:</u> Documents of the project must be submitted.

<u>Say:</u> You must submit Documents of your project.

Minimize use of abbreviations: An abbreviation is a condensed form of a word or phrase. Abbreviations may differ from place to place, region to region, or country to country. Unfamiliar abbreviations could create trouble for the reader in understanding the text. The reader constantly requires looking back to earlier pages or an appendix. In case everyone knows an abbreviation, you can use it without defining it. Examples are Dr., Mr., PhD, etc.

Sometimes, it is necessary to use abbreviations. In that case, the best practice is to define an abbreviation for the first time you use, for example, "International Organization for Standardization" (ISO). You should limit the number of abbreviations in the document to avoid confusion for the reader. Your aim is to make the document user-friendly.

Avoiding common vocabulary and spelling errors: What you write exhibits the quality of your communication skills and your attention towards the work. Using the wrong vocabulary or incorrect spelling will not effectively communicate your message to readers. Homophones in English sometimes create confusion while using the word. There are a few examples that commonly confuse most of us as follows.

<u>Advice:</u> a formal notice of a financial transaction.

<u>Advise:</u> offer suggestions about the best course of action to someone.

<u>Affect:</u> make a difference to

<u>Effect:</u> a change that is a result or consequence of an action or other cause

<u>Assure:</u> make (something) certain to happen

<u>Ensure:</u> make certain that (something) will occur

<u>Insure:</u> arrange for compensation in the event of damage to or loss of (property)

<u>Principal:</u> first in order of importance

<u>Principle:</u> a fundamental truth

<u>Stationary:</u> not moving or not intended to be moved

<u>Stationery:</u> writing and other office materials

<u>Than:</u> conjunction is used to introduce the second element in a comparison

<u>Then:</u> at that time

You can avoid these errors by proofreading the document, checking for the homophones, checking for single/double letters, and ensuring the presence of silent letters. Do not entirely rely on auto-correct or spell checker.

Use short, simple words - Replace difficult words and phrases with simpler alternatives: Word choice is important in making communication clear. When you have an opportunity to use simple words to explain the content, do not use pompous and complicated-sounding language. Using simple words will expand the scope the of audience for your document, ranging from beginners to advanced-level users. Gobbledygook (language that is meaningless or is made unintelligible by excessive use of technical terms) makes writing sound like a corporate voice rather than a person. Using gobbledygook will not help users understand or impress the audience; rather it complicates it.

Omit unnecessary words and repetition: Try to use simple phrases or statements. When you can explain the same statement using minimum and simple words, why use more words? Long, complex sentences containing many phrases cause confusion to the readers. When needless words are used in the statement, it becomes difficult to shape the statement and it becomes complex. Such complex statements would be difficult to understand by the users.

Extra words waste the reader's time. Remove information that the users don't need to know.

Check for "of", "to", "on", for, and other similar words. Here you can find an opportunity to reduce the phrase to one or two words.

Instead of "Do not hesitate to provide ...", you can use "Please provide ..."

Instead of "Prior to the juncture when ...", you can use "Before ..."

Instead of "on a yearly basis...", you can use "yearly..."

Remove redundant words from the statements.

Example statement with the redundant word: Describe your past medical history.

Example statement with removing the redundant word: Describe your medical history.

In the above example, the word "history" itself is self-explanatory and you need to describe the past about your medical conditions and medications. Hence, using the word "past" does not add any value and you can omit that.

Dealing with definitions: Use definitions rarely in the document. You can write a document in such a way that you can avoid needing to define a term. Definitions sometimes cause problems than they solve them. Follow the general rules which I consider useful while dealing with the definitions.

Do not define commonly used words in the definition section.

Define the term when it is used for the first time in the document.

If you have a definition section as a mandatory requirement, use it at the start of the document or chapter or at the end.

Only define the words that are used in the document.

Consistency in using words in the document: When you use different words or terms to say the same thing in the document, it often confuses the readers. For example, you need to describe a person, Ms. Rose John in the document. Sometimes you have used the name, Rose and sometimes Ms. John in the document. Using different styles may wonder readers if you are referring to the same person or if there are two different persons.

Avoid jargon, legal words, and pomposity: Jargon is special words or terminology related to a particular field. The jargon is only understood when used in a particular context by those who are members of a group or who are involved in a specific business. In other words, it is unnecessarily complex, technical words used to impress, rather than to inform, your readers.

Avoiding jargon does not mean leaving out necessary technical terms from the document. The first is the necessary use of a technical term in technical writing.

Jargon: Drill down

Simple language: To look at a problem in detail

Jargon: Involuntarily undomiciled

Simple language: Homeless

Legal terms used outside of legal audiences annoy them. Nobody wants to participate in discussions where they can't understand the language. People do not like when someone uses confusing or indirect language when simple and direct language is available. Jargon often complicates what is being said rather than promoting clarity, and often unnecessarily so. Following are a few examples, which can be avoided.

Aforementioned

Foregoing

Thereof

Wherein

Whereof

Hereby

Following is an example of complex and simple statements

"The said English book..." can be written as "The English book..."

Formatting

Formatting refers to the visual organization of a document on the page. The layout and formatting play a crucial role in making complex information reader-friendly and fulfilling the purpose of the content. A well-designed document guides the reader effortlessly to the required information, presenting it in a structured and organized manner.

When designing the layout and formatting of the content, always keep your audience in mind. For instance, if you're working on a project where you need to submit a report to a regulatory agency, the report should look formal, align with regulatory guidance, and simultaneously represent the organization's brand.

On the other hand, if you're working on training materials for a new employee in the organization, it would be more beneficial to include pictorials, diagrams, and infographics rather than a material full of text.

Sometimes, a writer may not have much choice if the organization has finalized a template to create a technical document in a specific format. This can save a lot of time, but the downside is that it doesn't allow the writer to show creativity.

When you start formatting, it's a good practice to identify a suitable audience and subject matter expert within the organization to get feedback about the formatting and the look's adequacy.

Using an authoring tool such as a style sheet is a sophisticated, abstract way to separate the formatting from the associated text. Using styles, the formatting can be stored, modified, and applied separately from the text. Make changes to the formatting once and let the styles do the work for you. If you're a beginner or not familiar with the tool, use the default template of a word processor or one designed by the organization.

Ensure that all the essential components of the document, such as the title, index, header, footer, and page number, are included in the document.

Document formatting can be improved using simple but very effective techniques such as bullets, charts, pictorials, tables, and callouts (Tips, Notes, Insights, etc.)

A callout is a short text defining a feature of a pictorial or schematic diagram, which provides information about that specific thing in the pictorial. The term is also used to describe related information about the topic in the form of tips, notes, and insights about the topic to attract attention. These are related pieces of information to the topic but not necessarily in the flow of text.

Unlike the content, the design of the content and graphics play an equally important role in grabbing the attention of the audience and reader. Content and design go hand in hand. Design instantly grabs the attention of the reader and impacts their opinion, while content has more of a slow, cumulative effect.

To have effective graphic design, there is a design principle developed by Robin Patricia Williams. The principle, known as CRAP, stands for Contrast, Repetition, Alignment, and Proximity. You can't just throw something together to create a design; a systematic approach and principles applied to the elements are what make or break a design. It's important to note that nothing is placed on the page by accident or coincidentally. By understanding CRAP, you can consistently deliver effective design, whether it's for a book, presentation, website, a landing page, or just a visiting card. Let's understand CRAP in brief with a hypothetical example of an invitation card.

In this example, I have selected an unformatted invitation card. The card has all the essentials that an invitation card should have. However, it is not adequately designed; hence, it's hard to distinguish different elements on the card.

You're Invited
The City Library
A community project 10 years in making
August 11, 2021
201, Green Leaf Avenue
Toronto
Canada
thelibrary@gmail.com
+112233445566

Contrast: Contrast is about making specific items easily visible and creating visual interest by applying dissimilar properties to the elements. This includes color, typefaces, shapes, size, texture, space, etc. Implementing contrast in design helps draw attention to important elements or create a hierarchy in the content, such as when one makes a heading's typeface point size larger than a subheading's. Creating contrast between the heading, title, and text will help guide the user to the important elements. We use contrast to create a focal point. Without contrast, an image or text is uneventful and the main idea gets lost.

By using this contrast principle of design, different colors, and font sizes can be used to make different components of the design stand out. Now, key information becomes much more distinguishable. This approach enhances the readability and overall user experience of the document.

The City Library

A community project 10 years in making

201, Green Leaf Avenue
Toronto
Canada
thelibrary@gmail.com
+112233445566

Repetition: Repetition of elements such as logos, colors, specific images, text, and font faces sets a thematic tone and appearance for the design. Repeating elements enhance visibility and provide a sense of consistency, representing the theme of brand recognition. The repetition of elements also establishes connectivity between them and helps to establish a correlation. Maintaining a balance between contrast and repetition creates a focal point.

By applying this principle of repetition in design, changing the color of the logo to match the name of the library, and placing the logo in the background, you've established a theme within the document. These repetitions of color and logo create a cohesive and visually appealing design. This approach enhances the overall user experience of the document.

Alignment: Alignment is indeed a crucial aspect of design. Nothing should be placed on the page arbitrarily; every element should have a visual relationship with the other elements on the page. Alignment plays a pivotal role in how the elements appear relative to each other, to the page, diagram, photograph, column, table, or even within a cell. It's a formatting attribute that determines the appearance of the design and aids in organizing and ordering the content. Aligning headings, text, and other objects of the content to the same imaginary line, whether you choose to align left, right, or center, can significantly enhance the readability and aesthetic appeal of the document.

The adequacy of alignment contributes to a neat design. In the given example, using center alignment for the logo and varying the text alignment (left, center, and right) to distinguish different contents is an effective strategy. This approach enhances the overall user experience of the document by making it visually organized and easy to navigate.

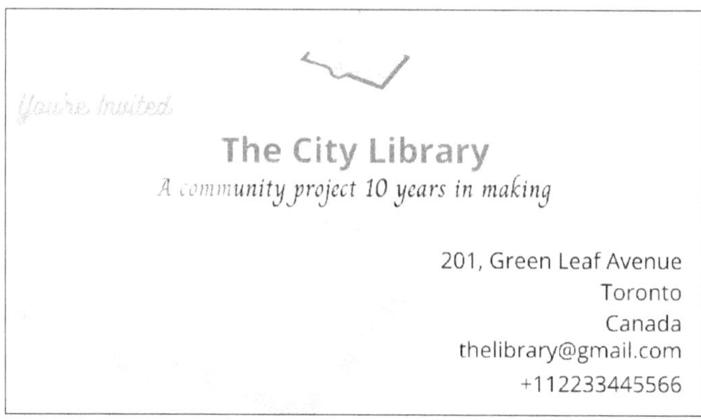

Proximity: The concept of proximity plays a significant role in design. It involves grouping related items together and maintaining a consistent distribution of space between objects. Consistent proximity in spacing and margins reinforces alignment and contributes to a neat design. Closer proximity helps establish relationships between objects and elements, making the design more coherent and intuitive. By effectively using proximity, you can guide the reader's eye and attention, enhancing the overall user experience of the document. This approach ensures that the design is not only visually appealing but also functionally effective.

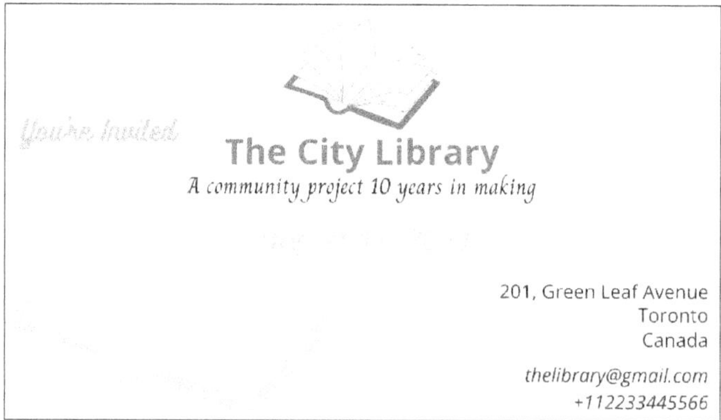

To make the card content more meaningful using the principle of proximity, I've created 5 groups.

Group 1: logo, name, and slogan.

Group 2: Invitation

Group 3: Invitation date

Group 4: Address

Group 5: Contact details

Related information within the group is in close proximity, which establishes a strong relationship between the content. This approach not only enhances the visual appeal of the card but also improves its readability, making it easier for the reader to understand and absorb the information.

	Draft initial content without worrying about spelling, grammar, or language initially. Focus on converting thoughts, creative ideas, and outlines into paragraphs to build the foundation. Embrace a collaborative approach in technical writing. Define roles such as author, editor, reviewer, and SME to different team members for efficient document creation. Plan for intermittent reviews with SMEs to gather feedback on content accuracy, flow, and practicability. Integrate examples, references, diagrams, and tips to enhance the usability of your technical content. Emphasize plain language and effective communication. Write in clear, direct, and

	plain language suitable for your audience's knowledge level. Tailor formatting to your audience and purpose. Consider audience expectations, whether it's a formal report or training materials, and use formatting to enhance document accessibility. Apply CRAP design principles (Contrast, Repetition, Alignment, Proximity) for effective graphic design. Ensure that design elements complement content, enhancing visual appeal.
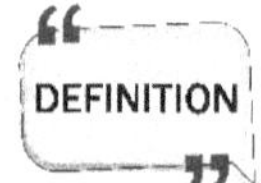	*Cyclical process:* A cyclical process is one in which a series of events happens again and again in the same sequence. *Callout:* A callout is a short text defining a feature of a pictorial or schematic diagram, which provides information about that specific thing in the pictorial. *Futz:* To waste time or tinker with something without a clear purpose or to little effect.
	*Summary:* Technical writing improves with iteration. Drafting is the heart, starting with groundwork like research and planning. You see, it's like we're on a team. Our subject matter experts (SMEs) subject matter experts (SMEs) sharing their knowledge and reviewers refining the narrative.

<table>
<tr><td></td><td>

Don't aim for perfection in the first draft, focus on getting the content down. Use visuals strategically and review for accuracy and audience clarity.

Formatting and design matter, so keep CRAP (Contrast, Repetition, Alignment, Proximity) in mind. And here's a little secret - the key to great communication is keeping it simple and clear. So, welcome those revisions with open arms, and before you know it, you'll be whipping up impressive technical documents!

</td></tr>
</table>

Chapter 7: Review

"The best way to find flaws in your writing is to have someone else look at it. A fresh set of eyes can catch mistakes and offer valuable suggestions."

~Stephen King

The review is a key aspect of the technical writing process. It's where every team member reviews their peers' drafts in a critical yet constructive manner, always considering the audience's needs and the purpose of the document. Expectations from the reviewer are to examine everything from the overall organization to the clarity of each paragraph, offering advice to enhance the writer's work.

When the final draft is done and has everything it needs, the review starts. This step is possibly the most crucial in technical writing. Before the document is submitted for review, the technical writer should evaluate its technical correctness through self-check and editing. A document filled with careless errors reflects a lack of seriousness on the writer's part and sends a negative message. Submitting a document with errors in the content is a grave mistake that undermines the review process.

When reviewers have to correct obvious and simple errors, they lose valuable time that could be spent focusing on more significant issues. If the writer eliminates these obvious errors, reviewers can concentrate on the more critical ones, adding value to the document.

The self-review should cover verification of document structure, layout, sentence structure, grammar, punctuation, and style.

It also focuses on content accuracy, ensuring the document's purpose is met, and assessing user-friendliness. This can be achieved by asking the following fundamental questions:

- Does the document provide the required technically important information?

- Is the document capable of helping users solve their problems?

- Is the write-up fact-based, and can evidence and references be provided?

- Is the information scientifically sound and meaningful to the users?

While technical reviews mainly focus on verifying technical content, editorial errors are not always easy to separate from them. A minor editorial error while writing terminology could change the meaning or technical aspect of the content. For instance, during chemical synthesis, writing the amount of solvent as mL (Millilitre) instead of L (Litre) could ruin the synthesis. So, it's essential to be mindful of such details.

Evaluating Content: Peer Review and SMEs

The review process can be divided into two parts. The first part involves a peer review within the department or function by experts in the relevant discipline. This peer review, conducted by fellow technical writers within the organization, helps identify writing issues, and technical errors and offers suggestions for document improvement. It also ensures compliance with organizational guidelines or regulations within the field.

Selecting a reviewer requires careful consideration. The ideal reviewer is an expert in the field, upholds high-quality standards, has extensive experience, and possesses subject knowledge.

Even though the reviewer is an expert, it's crucial to provide them with sufficient information and context about the work. This enables them to offer meaningful comments and review suggestions. Reviewers will keep in mind the expectations discussed with them. Friendly reminders about these expectations can be effective. It's a best practice to provide your expectations in writing so that the reviewer can refer to them as needed and track each one.

Another approach is a one-on-one document review. When a reviewer is extremely busy and hard to find time, discussing the document may be the only way to get valuable reviews. As a writer, you can employ this approach when working with engineers or field staff, especially when it allows you to demonstrate usability problems that may not be identified during a document review.

Provide expected time. Every project has a deadline. Ensure that you allocate sufficient time to address the comments and validate them before the final compilation.

Reviewers should consider a few important notes while providing review comments. The comments could be based on the reviewer's personal opinion, the organization's policy, or in reference to regulations or established standards. When review comments are opinions, they should preferably be called suggestions. The writer may or may not adopt these suggestions, depending on their judgment of how well the suggestion fits into the document.

Reviewers should praise sections that are well-written. Comments should be written in a positive manner, not sarcastically. They should focus on opportunities for improvement and solutions rather than flaws.

While providing comments, ensure that they are specific and relevant to the topic, add value to the content, are based on a thorough review of the document, and reference guidelines whenever necessary. Comments or suggestions should be explained in such a way that the writer understands the thinking and rationale behind the given comment and why it's important to address it in the document.

Organizing the Review Process

The review of a technical document involves various steps. The organization of the review process is crucial to ensure a timely and effective review of the document, ultimately producing a quality document ready for publication. The review process may vary depending on the size of the documents, the number of stakeholders involved, and the organization's policy. However, the following can be considered as general considerations while managing the review process.

Identifying Review Team Members and Defining Each Reviewer's Role

Defining the review team is vital. When identifying reviewers, consider different types of expertise to ensure that the document is reviewed with a holistic approach. When you give a document to experts, you want to extract value from their time. For instance, consider a technical content reviewer, an editorial content reviewer, a reviewer who can help with respect to regulations and corporate guidelines, and stakeholders who have the entire context with respect to the objective of the project or technical document.

The review team may consist of Subject Matter Experts (SMEs), a shop floor team, production managers, a safety team, a regulatory/ legal team, engineers, quality assurance, and sometimes customer representatives when an organization's strategy is Business-to-Business service.

Sponsor Review

The next step is to review the content with the sponsor of the project and collect their feedback. The review of the document by the sponsor is crucial to get feedback on whether the document is meeting the objective of its intent.

The formal review process is the final step to cross and reach the publication step. This step is a formal go-ahead from the stakeholders and sponsor to publish the final work. Also, the formal review is a requirement for certain standards and regulations such as ISO 9001, pharmaceutical, and biopharmaceutical quality systems.

Management of the Review Process

All projects have defined timelines. To complete the project on time and with effective outcomes, the management of the review is essential.

The first step of the review management process is to build accountability. Once the review team is finalized, communicate to all reviewers about the following:

- Mode of receipt of the document for review

- Role of each reviewer

- Timeline for completion of the review and giving feedback and comments

- Guidelines for commenting on the document

- Stage gate review meeting schedule

Here are some tips to manage the review process effectively:

Provide Appropriate Content for Review: Some reviewers prefer to review the entire document in one go, while others may prefer smaller pieces. Ask reviewers what they prefer. As reviewers progress through the document, their efficiency in finding errors may be reduced. It's important to provide background information and highlight important topics on which you want them to focus. This will help them stay attentive and provide valuable feedback.

Reviewing the document in parallel is also a good technique. Nowadays, there are many platforms available, such as Google Docs, and Microsoft SharePoint, where multiple people can work at the same time. The writer can continue to write, and reviewers can review and provide their comments or edit the documents. This is also possible with paper-based reviews, where the writer can provide a hard copy of the document to the reviewers.

Provide Prerequisite and Contextual Information: When you want to draw the reviewer's special attention to any specific topic, you should provide the information when you submit the document for review. Even though you use techniques like highlighting the text or providing comments, there's a chance that the reviewer might miss the requirement. Therefore, it's a good practice to discuss your expectations with the reviewer.

In-Person Review: Reviews take a good amount of time from the busy schedule of reviewers, especially SMEs or technical experts. Building a relationship and discussing the document with the reviewer may be the only way to get good reviews. Using a one-on-one discussion approach will help the reviewer understand your document easily and speed up the review process.

Track Timelines: Priorities are dynamic in nature, and it's very likely that reviewers might miss the timeline as they struggle to meet their own. Friendly reminders or requests for review are the best way to keep the reviewer attentive to the deadline, ensuring you get feedback on time. It's human nature to tackle easy tasks first, so it's a good practice to provide the document in parts and get it reviewed from time to time.

Offer support to the reviewer that might help them speed up the review process.

Appreciate Reviewers: Who doesn't like kind words? In the corporate world, most of us forget to say "thanks". Noticeable appreciation often produces exceptional results.

Managing Review Comments and Tracking:

Once the document is circulated for review, the technical writer will receive comments from the review team members. It's crucial to track these comments so that the writer can address them appropriately. The review process is cyclical, and there may be several rounds of review before the document is finalized and approved for publishing.

There are several ways that can be adapted to receive review comments and track those. It can be either or in combination of the following:

- Receiving comments as a hard copy

- Receipt of review comments in a word processor such as Microsoft Word in track change mode

- Receipt of review comments in a review sheet

- Use of a collaborative platform such as Google Docs where all reviewers can work at the same time and provide comments

- Documenting verbal feedback during review meetings

It's a good habit to compile all comments in one place if comments are received as a hard copy, in track change mode in a word processor separately, or comments noted during review meetings. While using a collaborative platform such as Google Docs, all comments are available in the same document and can be tracked easily.

In many organizations, for the formal review process, controlled review sheets are used in which the reviewer can provide comments, and it is retained for future reference as part of the documentation practice.

The technical writer should review all the comments carefully and address them appropriately. If the writer thinks that a comment is not required to be addressed, it should be explained and agreed upon with the reviewer why the writer does not need to address that specific comment. The writer should not take criticizing comments personally. If required, the writer should discuss review comments with the reviewer for a better understanding of the context to address the comment appropriately.

Once the received comments are addressed, it's a good practice to circulate the updated draft so that the reviewer can cross-verify that mutually agreed comments are addressed and that the revised version is appropriate and meets the objective. While circulating the updated draft versions of documents, the best way to track that the comments are being received in updated draft versions is to prevent any confusion while addressing them.

The Review Check Sheet

The review check sheet is an excellent tool for a structured review process. It helps the reviewer touch on every aspect of the document, capturing all minute details. It serves as a control mechanism to prevent failures that could occur due to minor errors.

The review check sheet should cover all essential checkpoints that are necessary for the document. These checkpoints should encompass the requirements expected by an organization, adherence to the organization's policy, good documentation practice, regulatory requirements, and checkpoints that ensure the objective of creating the document is fulfilled.

Here are some general checkpoints that could be covered as part of the check sheet:

- **Appropriateness of Document Format**: Ensure that the document follows the correct format.

- **Document Number and Version Number**: Check that the document number and version number are correct.

- **Good Documentation Practice**: This includes correct pagination, table of contents, correctness of headings, and adherence to the template, if any.

- **Correctness of References**: Verify that all references in the document are correct.

- **Correctness of Spelling, Grammar, and Punctuation**: Ensure that the document is free from spelling, grammar, and punctuation errors.

- **Correct Abbreviations and Capitalization**: Check that all abbreviations are correct and that capitalization rules have been followed.

- **Page Breaks**: Ensure that there are no page breaks that leave widows or orphans.

- **Check Tables, Graphics, and Images for Appropriate Placement**: Verify that all tables, graphics, and images are appropriately placed in the document.

- **Ensuring that the Objective of the Document is Meeting**: Check that the document meets its intended objective.

- **Review of Audience Needs**: Ensure that the document caters to the needs of the intended audience.

- **Layout and Design of Document**: Check that the layout and design of the document are appropriate and user-friendly.

- **Completeness and Correctness of Technical Content:** Ensure that the technical content in the document is complete and correct.

Ensure your document is technically sound by conducting a thorough self-review before you submit it to the reviewer. Eliminate careless errors to demonstrate professionalism and save reviewers time.

Choose reviewers carefully, prioritizing expertise, experience, and subject knowledge. Clearly communicate expectations in writing, enabling reviewers to focus on critical issues. Remind them of these expectations to enhance the review process.

When providing feedback, balance positive remarks with constructive criticism. Praise well-written sections and offer solutions for improvement. This approach fosters a positive environment and encourages continuous improvement.

Efficiently manage the review process by identifying team members, defining roles, and setting clear expectations. Tailor the review team to include various expertise types, such as technical content reviewers, editorial reviewers, and stakeholders. Conduct sponsor reviews to align the document with its intended objectives.

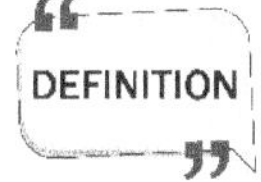

Widows and Orphans:

A widow is a paragraph-ending line that falls at the beginning of the following page or column, thus separated from the rest of the text. This leaves it by itself at the top of a page, with the rest of the paragraph on the previous page.

An orphan is a paragraph-opening line that appears by itself at the bottom of a page or column, thus separated from the rest of the text.

Summary:

Becoming good at technical writing needs one important step: the review process. The collaborative task between checking your own work and getting feedback from others will improve your work, making sure it's correct, clear, and useful for your readers. Think about skilled editors and experienced peers making your writing better, making it stand out for your audience.

Welcome the learning opportunity that feedback offers, and remember, reviewers are your helpers, not critics. Valuing the

	reviewers' feedback from reviewers can help you transform your drafts into polished, ready-to-use documents for your target readers. This collaborative review process will not only enhance your writing skills but also provide insight into the reviewers' perspectives.

Chapter 8: Edit

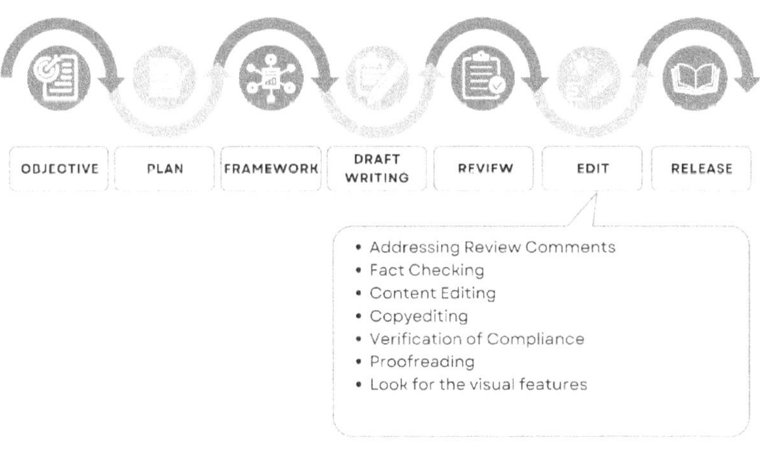

Editing is the process of polishing a document, making it error-free, and enhancing its readability. It's like crafting a masterpiece from a block of text, smoothing out the rough parts until it's perfect. It's a gatekeeping process where information is meticulously scrutinized before it's deemed fit for publication. Before a book is given the green signal for final publication, it undergoes multiple rounds of rigorous editing.

As writers, we pour our hearts and souls into our work, investing significant time and effort. It can be tough to cut out details that you've spent a lot of time researching and understanding, especially when they're interesting but not quite relevant. However, it's crucial to remember that any information that doesn't serve the reader or the context of the book must be trimmed.

While reading this chapter you might be thinking, hey, isn't reviewing and editing the same thing?" Well, let me tell you, they're different. Imagine you're a jeweler with a raw diamond in your hand. The first thing you do? You examine it closely, looking for any imperfections, assessing its strengths, and figuring out what needs to be improved. That's your review process. Now, once you've got a clear idea of what you're working with, you move on to the next step: editing. This is where you take your tools and start to work on that diamond. You chip away at the rough edges, polish it until it sparkles, enhancing its beauty and value. That's editing for you. So, while they might seem similar, reviewing and editing each play a unique and crucial role in the process of crafting a top-notch technical document. It's like turning a raw diamond into a dazzling gem.

The approach to editing a document can vary based on the editor's personal style and focus areas. In this chapter, I'll walk you through several key steps that are integral to the editing process. While some aspects of editing may overlap across these steps, each one plays a unique role in enhancing the quality of the document.

Step 1: Addressing Review Comments: This is where we take into account the feedback and suggestions from reviewers.

Step 2: Fact Checking: Ensuring the accuracy of the information presented in the document.

Step 3: Content Editing: This involves refining the content for clarity, coherence, and conciseness.

Step 4: Copyediting: Here, we focus on grammar, punctuation, and syntax to ensure the document is error-free.

Step 5: Verification of Compliance: Checking if the document adheres to the prescribed guidelines and standards.

Step 6: Proofreading: A final review to catch any overlooked errors or inconsistencies.

Step 7: Look for the visual features: Ensuring that the visual elements of the document, such as images, tables, and graphs, are appropriate and effectively presented.

Remember, editing is not just about correcting errors; it's about making your document the best version of itself. So, let's dive in and explore each of these steps in detail.

Addressing Review Comments

This step is crucial as it addresses the majority of technical improvements suggested by Subject Matter Experts (SMEs) and stakeholders. It's here that we tackle all the value additions, editorial errors, spelling mistakes, grammatical errors, and other obvious errors identified by reviewers. It's a step that sets the stage for the subsequent stages of editing.

Fact Checking

Fact checking involves verifying the content against specific references, guidelines, and policies of an organization, as applicable. It also includes verifying technical facts. If the document is a procedure or instruction manual, it's essential to verify the unit operation with respect to the defined process or steps in the document. This ensures that the written procedure is not just theoretically sound, but also practically applicable.

Content Editing

Also known as substantive or structural editing, this step focuses on the logical flow of content, structure, organization of content and visuals, and overall presentation. When editing services are outsourced, the editors start by reviewing the entire manuscript to understand the document's objective. Once they have a clear understanding, it becomes easier to organize and structure the content in a manner that's engaging and easy to follow.

Copyediting

Copyediting, also known as line editing, focuses on spelling, capitalization, grammar, and tailoring the language to the target audience. It's essential to ensure that the text aligns with the organization's policy. The words and phrases used in the document are checked for consistency across the document. Editors may refer to the organization's style books for guidance on sentence length, punctuation, or word usage if one is available. A general rule of thumb is to limit sentences to 18 to 20 words. Adhering to this rule makes sentences simpler and easier to read and understand.

Verification of Compliance

This step involves verifying the document's compliance with the organization's policies, legal requirements such as safety rules, and the use of copyrights, trademarks, and symbols. The best approach to this step is to develop a checklist. This ensures that no vital information is overlooked.

Proofreading

Proofreading is the process of meticulously reviewing the final manuscript to ensure its accuracy. It involves checking spelling, grammar, capitalization, consistency of words, abbreviations, and acronyms. The proofreader's role is to inspect and analyze the document for any errors that may have been overlooked by the writer or editor. While proofreaders generally do not make changes to the document, they highlight potential errors, leaving the final decision on corrections to the writer or editor.

Look for the visual features

Visual features such as diagrams, photographs, tables, graphs, symbols, maps, illustrations, schematics, and infographics not only draw the reader's attention but also aid in understanding ideas effortlessly. As the saying goes, "A picture is worth a thousand words." Visuals complement the written content and simplify the complex textual information.

It's essential to select relevant visuals that convey the message effectively. Poorly designed visuals can confuse the reader and have a negative impact. Here are some key rules for incorporating visual elements into the document:

1. Provide a clear title for each visual that can be referred to in the document.

2. Clearly label visuals. For example, parts of schematic diagrams, legends of engineering drawings or charts, X and Y axes of graphs, etc.

3. Credit the source of the information.

4. Maintain symmetry of images and graphs.

Visuals do not only cover graphical representation but also the layout of the document and formatting. This includes elements such as verification of the correct template, linking of text with the table of content, and use of the organization's correct logo.

Remember, each step in the editing process plays a crucial role in creating a clear, concise, and compelling document.

<table>
<tr>
<td>

</td>
<td>

The editing process is very important and should be done very diligently. Consider input from Subject Matter Experts (SMEs) and stakeholders seriously, tackling issues like value additions, editorial errors, spelling, and grammatical mistakes. This sets the foundation for subsequent editing stages, ensuring technical improvements are integrated.

Ensure accuracy by fact-checking against references and organization guidelines. For technical documents, verify that procedures align with defined processes. This guarantees not only theoretical correctness but also practical applicability.

Editing should be done considering requirements in the enhancement of the logical flow, organization, and overall presentation.

</td>
</tr>
</table>

<table>
<tr><td></td><td>Confirm adherence to organization policies and legal requirements. During proofreading, meticulously review visuals such as diagrams and charts. Follow best practices for incorporating visuals, providing clear titles, labels, and source credits.</td></tr>
<tr><td>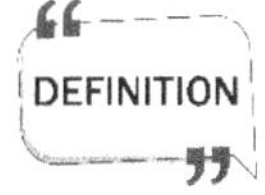
</td><td>

Gatekeeping:

The meaning of Gatekeeping refers to the process of controlling access or determining what is allowed and what is not. Here it is used metaphorically to describe the meticulous scrutiny of information before it's deemed fit for publication.

Conciseness:

Conciseness refers to the quality of being brief and to the point. In this chapter, it's mentioned about content editing, where the aim is to refine the content for clarity, coherence, and conciseness.

Tailoring:

In the context of copyediting, the term is used to describe the process of adjusting the language to suit the target audience. Copyediting involves checking spelling, capitalization, and grammar, and tailoring the language to the intended readership.</td></tr>
<tr><td>
</td><td>

Summary:

Editing is the magic touch that transforms rough writing into a gleaming gem. Editing is like polishing a diamond. First, you address reviewer feedback, incorporating suggestions and smoothing away errors. Second is fact-checking, which ensures the</td></tr>
</table>

<table>
<tr><td></td><td>

accuracy of the content and ensures that it's practical, not just theoretical. Next comes content editing, where you organize information logically, ensuring a smooth flow that captures your reader's attention.

Copyediting polishes the language, making it clear, concise, and aligned with your audience and style guide. Compliance verification is also an important aspect that's required during the editing process. It ensures that the document meets all regulatory and policy requirements.

Last but not least, it comes to proofreading. At this step, any remaining errors are caught through meticulous proofreading, and it assures that the final manuscript is flawless.

Remember, editing isn't just about words; visuals like diagrams and graphs play a crucial role too. Use visuals strategically. They should have clear titles and labels to enhance understanding and engagement for readers.

</td></tr>
</table>

Chapter 9: Release

"The true power of publishing lies in the ability to transcend boundaries, as words written by one can be read by many, connecting hearts across time and space."

~ Gabriel Garcia Marquez

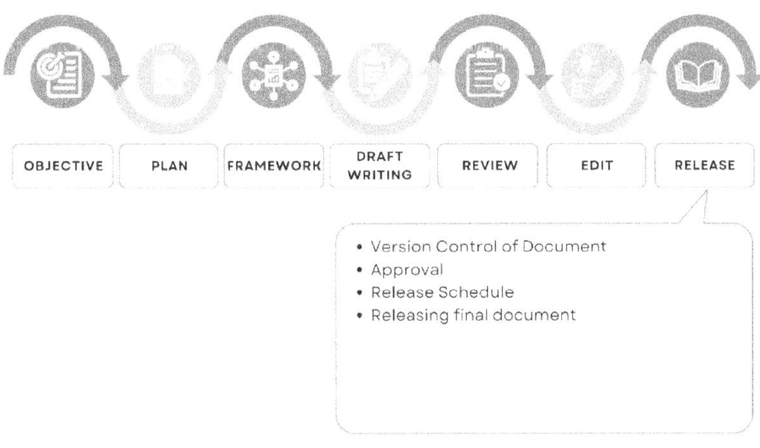

Releasing or publishing the final document to its intended audience is a significant milestone in the journey of technical writing. Once your manuscript has been thoroughly reviewed and edited, the final phase is to give it a final polish. It's crucial to ensure that your document is complete before it's handed over to the publisher or released through any electronic means.

Post the final edit, there should be no updates to the content or structure of the document. However, it's necessary to polish the document to ensure that it's ready for publication. This final touch-up includes quality checks, proofreading, and enhancing the visual appeals of the content and cover page. For proofreading, you can use grammar checkers or seek the help of a colleague to review the document.

You might be thinking about that again review? But wait, the final review of a technical document is essential to prevent the distribution of incorrect or confusing information, thereby enhancing the document's quality, effectiveness, and professionalism.

If your document is intended for print publication, it's a good idea to print a copy for review. A printed copy offers a different perspective and can help you spot things you might have missed on screen.

Moreover, a printed version provides a preview of the document, allowing you to check if the layout, styles, fonts, page breaks, headers, and footers are correct. It also helps ensure that the table of contents lists the correct headings and page numbers.

Depending on the document's significance and the level of error tolerance, you might consider employing a professional proofreader. This is typically preferred for documents that are difficult to revise post-publication, such as books, reports, user manuals, yearly reports, or commercially published publications.

Remember, a piece of writing may never feel fully finished to the writer. However, once you feel satisfied that you've addressed most of the questions and concerns, you've reached the end!

Version Control of Document

A document, once created, is often updated frequently based on changes to the product, new regulatory advice, or audience requests for improvements. Version control is a crucial tool that helps track when a document was created, revised, or released, and the reasons behind its various revisions.

To minimize the risk of using a document with an incorrect version number, the document version control process must be managed meticulously. Document version control allows for the tracking and management of multiple versions of a document. It's essential because it enables you to track changes, identify who made what changes, and if necessary, revert changes.

Here are the key components typically considered part of good documentation practice and for maintaining version control of documents:

Document Number: This is a unique identification number for the document, which can be numeric or alphanumeric.

Version Numbers: These are typically assigned based on the changes to be made. Here is an example of version control. The main version number is followed by a decimal-point increment for the minor version. For example, 0 to 0.1, 0.2, and 0.3. When there's a significant or major change, the version number increases to 0, 1, and so on. Version numbering is crucial as it helps distinguish between different versions of a text. A simple numbering scheme consisting of consecutive whole numbers may suffice to keep track of the version of a document you're working on. Such version numbers may be referred to as Version and Revision for minor and major changes respectively.

Minor Version Number: This refers to minor changes to a text, such as spelling or grammar corrections.

Major Version Number: Changes that necessitate a document's reapproval are considered major changes. The complete number is increased by one to indicate major changes.

Reason for Revisions: Keeping a record of the reason for each revision is important to track the changes made during each revision. This allows the organization to track the document's history, why it was revised, when it was revised, and the purpose behind the revision. This makes the document self-explanatory and eliminates the need for individuals to remember and explain over time.

Approvals: This section of the document control includes details of team members who have reviewed and approved the respective versions of the documents. It should include the name, job role or title, and department names.

Document Approval Date: This is the date when the document is approved for publication.

Document Effective Date: Some documents are approved before release and circulated to relevant stakeholders. This arrangement is made when stakeholders need time to prepare before the technical document becomes effective. As soon as the document is effective, stakeholders will be able to comply with the requirement. Examples include regulatory norms, industry best practice guidelines, or any release of software updates.

Approval

The approval process is a critical step in the document lifecycle. It involves a thorough review by key stakeholders who have the authority and expertise to validate the document's content. This process ensures that the document is accurate, complete, and adheres to the organization's standards and guidelines. The approval is typically documented with the approver's name, title, and the date of approval. It's important to note that the approval process may vary depending on the organization's policies and the nature of the document. However, the ultimate goal remains the same: to ensure that the document is ready for release and meets the intended purpose.

Release Schedule

The release schedule outlines the timeline for when the document will be made available to its intended audience. This schedule is carefully planned to align with product launches, regulatory deadlines, or other key milestones. The release schedule should consider the time required for final approvals, printing (if applicable), and distribution. It's crucial to communicate the release schedule to all relevant stakeholders to ensure a smooth and timely release. Remember, a well-planned release schedule can greatly enhance the impact and reception of your document.

Releasing final document

Once the document receives the green light, it's time to roll it out as per schedule. There are various avenues to release the final document. Here are a few examples:

- The traditional method of publishing a bound book
- Technical articles published in scientific journals
- Posting on the organization's web portal
- Circulating the document to stakeholders via email

After the document goes live, it's best practice to make an announcement. The mode of communication depends on the type of document you've released. It could be through the company's intranet or an email to relevant stakeholders. In today's digital age, professional social media platforms like LinkedIn also serve as excellent channels for public announcements to relevant professional groups and communities.

Communication, though a small part of the entire task, is one of the critical aspects. The right message to the right audience is the key to a successful release of a technical document. Before sending out the communication, run the draft by the relevant stakeholders, colleagues, and key individuals in the organization who are responsible and accountable for the document's release. Their input can be invaluable in ensuring that communication is effective and well-received.

| 
QUICK
TIPS! | Ensure a meticulous final review before publication to prevent the dissemination of incorrect or confusing information. Despite potential reluctance, this step enhances the document's quality and professionalism. Utilize proofreading tools and consider seeking input from colleagues or professional proofreaders.

For documents intended for print, print a copy for review. A printed version provides a different perspective, helping identify potential issues with layout, styles, fonts, and other visual elements that may go unnoticed on screen. |

Implement a robust version control system, including unique document numbers, version numbers, and reasons for revisions. Clearly define minor and major changes, record approvals, and track the document's history. This ensures accuracy and clarity, helping avoid the use of incorrect versions.

Plan a well-thought-out release schedule aligned with product launches, regulatory deadlines, or other milestones. Communicate the schedule to relevant stakeholders and consider factors like final approvals, printing, and distribution time. A well-planned release schedule enhances the document's impact and reception.

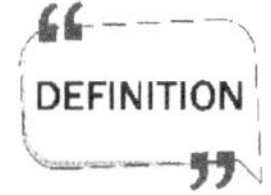

Dissemination:

Dissemination refers to the act of spreading or circulating information widely.

Meticulously:

Meticulously means with great attention to detail and thoroughness.

Avenues:

Avenues refer to different channels or methods through which the document can be released, such as publishing a bound book, posting on a web portal, or circulating via email.

Intranet:

An intranet is a private network within an organization that uses Internet technologies to share information, collaboration tools, and operational systems among employees.

	Green Light: "Green light" is a metaphorical expression indicating approval or permission to proceed with the release.
	*Summary:* Polishing your technical masterpiece for release is like adding that final flourish to a delicious cake. After meticulous edits and reviews, give it a final proofread and visual touch-up. Track revisions with version control, like labeling cake layers, and ensure approvals by key stakeholders, like getting a thumbs-up from a professional judge. Plan the release like a birthday party, with a schedule for printing and distribution. Share your creation through print, online portals, or even social media, just like showcasing your cake at a bake sale or on Instagram. Remember, clear communication is the icing on the cake, so inform your audience through the right channels and let them savor your technical masterpiece!

Chapter 10: Conclusion

"The art of writing is the art of discovering what you believe."

Gustave Flaubert

Accomplishing a technical document and releasing it to the intended audience gives enormous pleasure to the author and the entire team. It's a time to pat yourself on the back and share a round of applause with your team.

The goal of technical writing is to distill complex information into a form that even readers with less technical knowledge can grasp. As an author, it's crucial to remember that your reader may be new to the topic or only have a basic understanding of it. Crafting quality content with a technical focus might seem straightforward, but it requires some unique skills.

Technical writers serve as bridges, connecting experts to readers. They write concise, factual information to educate and explain. If you're a technical writer, you need to have a solid understanding of the domain you're working in. But that's not all. To broaden the scope and prospects of your writing, you also need to be adept at research, language, writing, and proofreading.

A well-crafted text is a joy to read. The information you need is easy to find, the document flows smoothly, the language is clear and concise, and the tone is professional.

You have reached the end of this book on the technical writing process. Congratulations on completing this journey of learning how to create clear, concise, and effective technical documents.

However, your work as a technical writer is not over yet. As you know, many technical documents require regular updates and maintenance to keep up with the changes in the products or processes they describe. Depending on the extent of these changes, you may need to repeat some or all of the steps of the technical writing process.

For example, if you are writing about a software that has frequent bug fixes or minor feature additions, you may only need to revise and publish your document to reflect these changes. But if you are writing about a product that has undergone a major overhaul or a process that has been redesigned, you may need to start from the beginning and plan your document again.

Updating a document can be challenging, especially if you have not touched it for a long time. You may need to re-familiarize yourself with the subject matter, the audience, and the purpose of the document. You may also need to do more research, analysis, and testing to ensure the accuracy and validity of your information.

Therefore, do not take updating a document lightly. It is an essential part of the technical writing process that ensures the quality and usefulness of your document. By updating your document regularly, you are providing a valuable service to your readers and stakeholders.

We hope that this book has given you the knowledge and skills you need to succeed as a technical writer. I also hope that you have enjoyed reading it as much as I have enjoyed writing it.

Thank you for choosing this book and happy writing!

A Request to The Reader

May I ask you for a small favor?

I would like to thank you from the bottom of my heart for having bought and read my book. I hope this book has met your expectations and given you good insight and knowledge regarding "Technical Writing Simplified".

I request you to provide me with your valuable rating and a review of the book on the Amazon site. Your feedback would inspire and encourage me in my author journey, where I will look forward to enhancing your knowledge, skills and positively impacting the lives of many professionals.

Please leave your review by scanning following QR Code, it will directly lead you to book review page.

Thank you once again. Wish you all success and happiness in your life.

References

Levy, M. (2010). *Accidental Genius: Revolutionize Your Thinking Through Private Writing*. National Geographic Books.

Hart, J. (2007). *A Writer's Coach: The Complete Guide to Writing Strategies That Work*. Anchor Books

Henneke. (n.d.). 3 Writing Strategies: How to Put Your Thoughts Into Words. Enchanting Marketing. Retrieved January 17, 2024, from Enchanting Marketing (https://www.enchantingmarketing.com/).

Plain Language Action and Information Network. (2011). *Federal Plain Language Guidelines, March 2011, Rev. 1, May 2011*. Retrieved January 17, 2024, from Plain Language (https://www.plainlanguage.gov/howto/guidelines/Federal PLGuidelines/FederalPLGuidelines.pdf).

Lirola, M. M. (2009). *Main Processes of Thematization and Postponement in English*. Peter Lang AG

Alred, G. J., Brusaw, C. T., & Oliu, W. E. (2012). *Handbook of Technical Writing*. Bedford/St. Martins

Elling, R., Andeweg, B., de Jong, J., & Swankhuisen, C. (2012). *Technical Writing Skills*. Translated by K. van der Linden. Institute of Technology and Communication, Delft University of Technology

Straus, J. (2011). *The Blue Book of Grammar and Punctuation: An Easy-to-Use Guide with Clear Rules, Real-World Examples, and Reproducible Quizzes (10th ed.)*. John Wiley & Sons

Colman, R. (2011) *The Briefest English Grammar and Punctuation Guide Ever!* University of New South Wales Press

Fenton, N. (2010). *Improving your Technical Writing Skills* (Version 6.0). School of Electronic Engineering and Computer Science, Queen Mary (University of London)

Williams, Robin, 1953-. (1994). *The non-designer's design book: Design and typographic principles for the visual novice*. Berkeley, CA: Peachpit Press

About the Author

Karim Panjwani, a pharmacist, quality assurance professional, and technical writer, brings over 17 years of experience in the pharmaceutical industry. He has collaborated with leading multinational companies to establish quality systems that adhere to various regulatory bodies, including the U.S. FDA, MHRA, EU GMP, Health Canada, TGA, ANVISA, and WHO Geneva.

His area of expertise encompasses oral solid dosage, sterile, and nasal formulations. He possesses a profound understanding of the manufacturing and utility equipment used in pharmaceutical formulation production. Karim is adept at drafting and editing technical documents such as procedures, processes, investigations, risk assessments, and regulatory communications. He has a comprehensive understanding and practical knowledge of global regulations concerning the manufacture, storage, and distribution of pharmaceutical medicines.

Karim is passionate about disseminating his knowledge and insights through his papers and blogs. These cover a range of topics, including good documentation practice, regulatory expectations, quality culture, and cleaning validation.